They Came to Locust Grove

The Saga of the Clark and Croghan Families
Who Influenced the Growth of
Kentucky and Our Nation

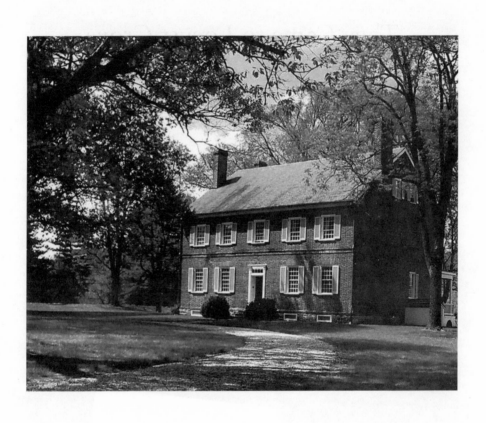

They Came to Locust Grove

*The Saga of the Clark and Croghan Families
Who Influenced the Growth of
Kentucky and Our Nation*

By Melzie Wilson

HARMONY HOUSE PUBLISHERS

Harmony House Publishers
P.O. Box 90
Prospect, KY 40059
502-228-2010

Design: Laura Lee

Printed in the USA

Hardback ISBN: 1-56469-117-9
Softback ISBN: 1-56469-121-7
Library of Congress Number: 2004114154

DEDICATION

⚜

In Memory Of
Husband John and Son Dale
Died February and March of 1988

⚜

Dedication
Neal, Elizabeth and Roger

Contents

Photographs ix

Preface xi

Acknowlegements xiv

Part I
The Clark and Croghan Families of Locust Grove

Chapter One   Clarks and Croghans .3

Chapter Two   On the River to Kentucky 11

Chapter Three   Mulberry Hill and Frontier Life 14

Chapter Four   The Courtly Major of Locust Grove 21

Chapter Five   Locust Grove 31

Chapter Six   Explorers Leave the Falls 44

Chapter Seven   General George Rogers Clark 59

Chapter Eight   1811 Earthquake and River Trade 70

Chapter Nine   War of 1812 78

Chapter Ten   A Day in the Life of Lucy 83

Chapter Eleven   Social Life—Matchmaking—Marriage 93

Chapter Twelve   Eventful Year for Lucy 106

Chapter Thirteen   Last Days at Locust Grove 114

PART II
Born to be a Soldier

Chapter One	The Livingstons	125
	Livingston Family Chart	
Chapter Two	British March on Livingston Manor	129
Chapter Three	Born to be a Soldier	137
Chapter Four	George and Serena	144
Chapter Five	New Orleans	149
Chapter Six	Inspector General on the Frontier	157
	Excerpts from: Official Inspection Reports Made by Colonel George Croghan 1826–1845	171
Chapter Seven	Mexican War	178
Chapter Eight	Serena and Tinie on Their Own	186
Chapter Nine	Civil War—Orphans—California	192

PART III
Mary Croghan and the Englishman

Chapter One	O'Haras of Pittsburgh	203
Chapter Two	Mary Croghan Had Two Families	214
Chapter Three	Romance That Rocked Two Continents	219
Chapter Four	Surinam—Dutch Guinea	226
Chapter Five	Schenley Children	239
Chapter Six	Mary Wins Over U.S. Courts	245

PART IV
Widow of Lafayette Square

Chapter One   At Home in Kentucky 257

Chapter Two   Beaus and Belles of Washington 264

Chapter Three   California 274

Chapter Four   Widow of Lafayette Square 293

PART V
The Croghan Heirs of Mammoth Cave

Family Charts 333

Notes 339

Bibliography 366

PHOTOGRAPHS

1. St. John's Church, Richmond Virginia, Library of Congress, p. 5.
2. Monument of President Thomas Jefferson's sister, Lucy Jefferson Lewis, Smithfield, Kentucky, p. 30.
3. Major William Croghan, 1820 by John Wesley Jarvis. Courtesy of Historic Locust Grove, Inc., p. 33.
4. Lucy Clark Croghan, 1820 By John Wesley Jarvis. Courtesy of Historic Locust Grove, Inc., p. 34.
5. John O'Fallon (Colonel) By George Calder Eichbaum, Missouri Historical Society, p. 37.
6. Benjamin O'Fallon (Major) By Chester Harding, Missouri Historical Society, p. 37.
7. Charles Thruston, Courtesy of Charles Timothy Todhunter, p. 38.
8. Frances Ann Thruston Ballard, Courtesy of Charles Timothy Todhunter, p. 41.
9. Lewis & Clark map of expedition, p. 47.
10. William Clark, By Charles Wilson Peale, 1810. Independence National Historical Park, Philadelphia, p. 52.
11. Julia Hancock Clark by John Wesley Jarvis, 1820, Missouri Historical Society, p. 53.
12. Harriet Kennerly Clark, By Chester Harding, From W. C. Kennerly, *Persimmon Hill*, 1948, p. 53.
13. General George Rogers Clark By Matthew H. Jouett. Courtesy of The Filson Club, Louisville, Ky., p. 60.
14. Locust Grove by John Nation, Courtesy of Historic Locust Grove, Inc., p. 84.
15. Eliza Croghan Hancock, Courtesy of Historic Locust Grove, Inc., p. 99.
16. Quartermaster General Thomas Sidney Jesup in White House Quarters (1788–1860). By Samuel King, Library of Congress, p. 102..
17. Dr. John Croghan, Purchased and developed Mammoth Cave Resort [1839]. Courtesy of Historic Locust Grove, Inc., p. 115.
18. William Clark Monument in Bellefontaine Cemetery, St. Louis, by Emil Boehl, ca. 1902, Courtesy of Missouri Historical Society, p. 120.
19. Eliza McEvers Livingston, [Serena Croghan's mother] By John Vanderlyn, 1802. Senate House paintings, New York State Office of Parks, p. 132.
20. Serena Livingston Croghan By Thomas Sully, 1815. Gift from Donald Newhall [Serena's Great Grandson Donald Newhall] Sisters Serena and Angelica have same harp in foreground, Historic Locust Grove, Inc., p. 134.
21. Angelica Livingston, By Thomas Sully, [Painting is signed "TS 1815"], p. 135. Angelica died in France, nineteen years of age, New York State—Bureau of Historic Sites.
22. Col. George Croghan, By John Wesley Jarvis. Courtesy of Historic Locust Grove, Inc., p. 138.
23. The Four Generations: 1913 or 14. Courtesy of John Wheaton. Top left, Virginia Rogers Nokes Murphy, B. 1882. Top center, Last heir of Mammoth Cave Serena Eliza Livingston Croghan Rodgers, B. 1832. Top right, Cornelia Livingston Rodgers Nokes, B. 1859. Front, Virginia Murphy Wheaton, Mother of John Wheaton, p. 190.
24. Christmas 1960, Jane Wyatt of movies and television fame. Seated left: Husband Edgar B. Ward died 2001 after 64 happy years of marriage. Sons, Christopher Ward and Michael Ward, p. 195.
25. Samuel Thruston Ballard. Courtesy of Charles Timothy Todhunter, p. 197.
26. Rogers Clark Ballard Thruston, Courtesy of Charles Timothy Todhunter, p. 197.
27. Memorial to Colonel George Croghan in Fremont Ohio, Courtesy of Ohio Historical Society, p. 200.
28. William Croghan, Jr., Courtesy of Historical Society of Western Pennsylvania, p. 206.

29. Mary O'Hara Croghan, Courtesy of Historical Society of Western Pennsylvania, p. 207.
30. Mary Croghan, Fourteen years old, student at school in New York, Courtesy of Carnegie Library of Pittsburgh, p. 220.
31. Captain Edward W.H. Schenley. The British army officer whose third elopement was with Pittsburgh Heiress Mary Croghan. He was 43, she 15. Two earlier wives had died. He won her heart, but didn't win her money. Pittsburgh's Schenley Park bears his name. Historical Society of Western PA., p. 221.
32. Mary Schenley in her later years. Historical Society of Western Pennsylvania, p. 240.
33. Lily Schenley Harbard, at Picnic, Pittsburgh PA., Historic Locust Grove, Inc., p. 241.
34. Jane Schenley, Pittsburgh (1855–1858), Historic Locust Grove, Inc., p. 242.
35. Mary Jesup Blair, in Nice, 1867–1868, Historical Society of Washington, D.C. p. 294.
36. Julia Jesup, in Nice, 1867–1868, Courtesy of Historical Locust Grove, Inc., p. 298.
37. Violet Blair Janin at Nice, 1867–1868, Historical Society of Washington, D.C., p. 299.
38. Jesup Blair, Courtesy of The Henry E. Huntington Library, California, p. 300.
39. Luti James (Jimmie) Blair, Courtesy of The Henry E. Huntington Library, California, p. 300.
40. Meriwether Lewis Clark, Jr., Historic Locust Grove, Inc., p. 301.
41. Meriwether Lewis Clark, Courtesy of Missouri Historical Society, p. 303.
42. Abigail [Churchill] Clark, Courtesy of Missouri Historical Society, p. 303.
43. Croghan Family Reunion: September 2002, Locust Grove, Louisville, Ky., p. 331.

PREFACE

Historically, history has been written about men, for men, and by men. These writings are of great value, but most often something is missing—the humanity, the persona, the flavor of everyday life, and glaringly omitted are women, children and family. This process or mind-set has often been carried over to most things involving history, including what is told during a tour at a historical home.

During my years as a docent at Historic Locust Grove, Inc. in Louisville, Kentucky, I found myself wanting to know more about the women of long ago whom I introduced to visitors by pointing to portraits. Visitors learned that the house, built in 1790, was restored in honor of General George Rogers Clark, Revolutionary War hero, conqueror of the West and founder of Louisville. Clark had spent the last nine years of his life with his sister Lucy Clark Croghan and her family at Locust Grove. At the end of a tour our guests would have acquired some knowledge of the other men in the family. First, there was the builder of the house, Major William Croghan, who fought with General George Washington at Brandywine and crossed the Delaware River under extreme conditions. Next, there was General Clark's younger brother, William Clark of the 1803–06 Lewis and Clark Expedition. Then, Major Croghan's son, Colonel George Croghan, who gained national fame for the defense of Fort Stevenson during the War of 1812, was very proudly brought to the attention of our guests.

In addition to the celebrated men, my interest began to grow as I realized women and family have been essential to the stability of households, culture, and communities. Those who performed the job as wife, mother and running a household had stories to tell about their joys, their struggles, their grief, their strengths, and weaknesses.

What about the women who had lived at and were connected to Locust Grove? Could Major Croghan have been a successful slave holding landowner without the help of his wife, Lucy? Much of the time, Lucy was in charge of Locust Grove while her husband was away surveying land grants and tending to his commercial trade on the Ohio River.

Over a period of time the picture of long ago began to change for me. A "Gentleman's Country Seat" required the ability to organize and administer an entire enterprise of this kind. A husband and wife were both necessary. Also, a reminder was: we wouldn't know about these women had it not been for the heroic men in their lives.

As time passed, I realized the Clarks and the Croghans were remarkable people. America owes so much to these families, with their ambitions and dreams, who became builders of the American frontier, the backbone of America between the East and West Coast. However, building America didn't stop with the frontier. This writing is a continuous long detailed account of five generations of remarkable Clarks and Croghans ending in early Twentieth Century.

While researching and writing, thoughts began to surface to the year 1784 when John and Ann Rogers Clark left Caroline County Virginia and came down the Ohio River. Did they dream their children and descendants would leave a claim to fame and greatness which is recognized and celebrated today? Monuments have been erected in their honor of these descendants. Kentucky's most noted landmarks, Mammoth Cave in Western Kentucky and Kentucky Derby at Churchill Downs, Louisville, Kentucky were established and developed by Clark grandsons. A Croghan granddaughter saved Mammoth Cave from falling into disrepair. Today, Pittsburgh owes much to a Croghan granddaughter who made it possible for her home town to have hospitals, museums and parks.

Visitors come from far and wide to visit Locust Grove, the last home of the old patriot, General George Rogers Clark. More than 200 years have passed since the stately Georgian style house on the Ohio River was built, but today, it stands as a proud monument to those who gave so much for their country.

My research in libraries, family letters and records has been rewarding. Some librarians I contacted knew very little about General George Rogers Clark and expressed a desire to learn more.

As a result, digging in archival collections for family letters has been my lot in life for the last twenty years. Looking for that piece of puzzle that had been missing for so long had become an addiction fed by elation when suddenly and unexpectedly the facts appear before my eyes. Windows were opened into their lives. Untouched letters left by the family and descendants enabled me to visualize the Croghans dancing in the ballroom with cousins and friends; budding romances; the anticipation of waiting for a loved one to return from a long absence; hearing the cry of a newborn baby, or the sounds of grief coming from the family graveyard not far from the house.

My love for history has always been present, but through the Clark and Croghan family letters, people, places and events of two hundred years ago, came alive. Human documents that I found moving, such as tear drop stains left on letters, have been included.

This book will serve as a monument, especially, to Lucy Clark Croghan, her daughter-in-law Serena Livingston Croghan and two granddaughters, Mary Jesup Blair and Mary Croghan Schenley as well as the heroes in the family. Out of four generations of women in the family, four were chosen as special people who bravely stepped forward and took a stand for a cause in what was a "man's world" in those days. They overcame disappointments, challenges and difficult times in their lives and also made things happen for the good of others.

My mission for writing this is to tell the story, allowing readers to experience close up the small events as well as great ones that helped shape the destiny of this family.

Melzie Wilson

ACKNOWLEDGEMENTS

Sometimes when researching, new evidence is discovered which discounts that which has been accepted as fact. Many family letters which have lain untouched in library archives for almost one hundred years are rich in new information. The following people are museum directors, archivists, librarians and researchers who were found to be very accommodating and helpful. The attention given to my many requests has not gone unnoticed.

Research performed many years ago by Samuel W. Thomas and Ms. Emilie Strong Smith at Historic Locust Grove, Inc., Louisville, has served as the foundation for writing this book. Those at the Library of Congress, I found to be very resourceful and courteous employees who were quick to respond and interested in my project. I want to thank the following:

Carnegie Library, Pittsburgh, Pennsylvania

Peter J. Blodgett and Brooke M. Black at Huntington Library, San Marino, California

Barbara Paff, Rutherford B. Hayes Presidential Center, Spiegel Grove, Fremont, Ohio

Suzanne Wiersma, The Birchard Public Library, Fremont, Ohio: Carolyn Bahnsen, Croghan Day Committee Chairman, Fremont, Ohio

The Historical Society of Washington, D.C.

The New York Historical Society, Joseph Ditta

The New York Public Library, John D. Stinson

Oregon Historical Society, Portland, Oregon, Shawna Gandy, Mikki Tint, Todd Welch and Lucy Kopp

Missouri Historical Society, St. Louis, Missouri, Ellen Thomasson, Amanda Claunch and Joseph Carlisle

National Archives and Records Administration. Washington, D.C., Trevor K. Plante

Kentucky Historical Society, Frankfort, Kentucky

LDS Family History Center Resources, Louisville, Kentucky and Clarksville, Indiana

Historical Society of Western Pennsylvania, Pittsburgh, Pennsylvania, Rebekah Johnston

The Historical Society Washington, D.C., Ryan M. Shepard, Bonnie Hedges, Curator of Collections, Gail Redmann

Princeton University, Princeton, New Jersey

Montgomery Historical Society, Rockville, Maryland

The National Society of The Colonial Dames of America, Washington, D.C., Amy D. Andrews

Ohio Historical Society, Columbus, Ohio, Duryea Kemp and John Haas

Columbia University, Washington, D.C.

Silver Spring Historical Society, Silver Spring, Maryland, Jerry McCoy

Charles Timothy Todhunter, author, Payneville, Kentucky

Those at The Historical Filson Club, especially Rebecca Rice and James Holmberg, Louisville, Kentucky

The National Society of the Sons of American Revolution, Louisville, Kentucky, Michael A. and Robin Christian

Yale University Library, New Haven, Connecticut

Independence National Historical Park Library, Philadelphia, Pennsylvania

Senate House State Historic Site, Kingston, New York, Joyce Kobasa and Deana Preston

Albemarle County Historical Society, Charlottesville, Virginia

To the following a very special thanks:

Jane Wyatt Ward, Los Angeles, California, John Wheaton, Sacramento, California and his mother Virginia Wheaton (1905–2000), George Rogers Clark Stuart, Abingdon, Virginia

Marge Nevin, Pee Wee Valley, Kentucky, Lois Bordner, Mary Ann and Joe Thorp, Louisville, Kentucky

Stephen M. Knowles, Falls of the Ohio State Park, Clarksville, Indiana

Jane Sarles, Clarksville Historical Society

PART I

The Clark and Croghan Families of Locust Grove

Breathes There the Man

From "The Lay of the Last Minstrel" CANTO VI

Breathes there the man with soul so dead
Who never to himself hath said,
This is my own, my native land!
Whose heart hath ne'er within him burned,
As home his footsteps he hath turned
From wandering on a foreign strand?
If such there breathe, go, mark him well;
For him no minstrel raptures swell;
High though his titles, proud his name,
Boundless his wealth as wish can claim,
Despite those titles, power, and pelf,
The wretch, concentred all in self,
Living, shall forfeit fair renown,
And, doubly dying, shall go down
To the vile dust from whence he sprung,
Unwept, unhonored, and unsung.

Sir Walter Scott

CHAPTER ONE

Clarks and Croghans

Between the years 1803 through 1822, Locust Grove was the most visited house in Louisville, Kentucky. No matter how wealthy Lucy Clark Croghan's husband was or how large the country seat Locust Grove, which sat on top of a range of hills overlooking the Ohio river six miles from Louisville, leisure time could not be bought. It took all her time to supervise the servants in cooking, sewing and laundering. She had to plan and carry out the work for her large family, not to mention the heavy demands of hospitality. Clark relatives and visitors were always present at Locust Grove. Lucy was expected to entertain graciously.

President Monroe, General Jackson and their retinue were weekend guests in 1819. Zachary Taylor grew up on the adjoining property called Springfield and until his death he visited Locust Grove quite frequently as if returning home.

Locust Grove is the only existing historic landmark where the hero General George Rogers Clark lived the last nine years of his life. He fought and won from the British the North West Frontier which gave the nation all land from the Ohio River to the Canadian border east of the Mississippi River. Over two hundred years have passed since then and today, visitors come from far and near to visit Locust Grove.

It was here that William Clark and Meriwether Lewis returned for a week of celebrating after the Lewis and Clark Expedition. Seven years later, the Croghan's twenty one year old son, Colonel George Croghan, the most celebrated hero of the War of 1812, returned home to Locust Grove.

Many books have been written about the two famous Clark brothers and their Croghan nephew. However, it was many years later before historians and publishers began to realize the real story of the winning of the West during the Revolutionary War.

Those living east of the Allegheny Mountains failed to take notice of the important achievements General George Rogers Clark had won from the British on the North Western Frontier. Americans today do not realize or appreciate how close the Ohio River was to becoming the Canadian border at the Peace Treaty of 1783. Britain was demanding the return of all the land north of the Ohio River, which Clark and his men had fought so hard for—they had given to the new country Wisconsin, Michigan, Illinois, Ohio and Indiana.

General George Rogers Clark was the hero who led his men through unbelievably bad conditions. Under his leadership and almost alone with a small group of pioneers, he won this northern empire for the new nation and saved the five southern states from the British.

In the winter of 1778, General Clark learned that "The Hair Buyer" British officer Henry Hamilton had plans for capturing lands below the Ohio River in the spring, and capturing all the territory south and west of the Alleghenies. Clark knew he could not wait for Hamilton's move. Only a surprise attack in the dead of winter would save the people in the west. Clark led his band of 175 men, in the month of February 1779, across Illinois, wading flooded lands, breaking ice to cross the swollen rivers. During the last three days, in desperate cold, they had almost no food. His surprise attack on Vincennes is one exciting event.

In this one brave stroke the land area doubled for the United States and the West was opened, carrying the American dream all the way to the Pacific.

The youngest Clark brother William was one of those responsible for carrying the American Dream to the Pacific Ocean and securing that claim.

Virginia was the birthplace of many patriots, heroes and nation builders. The following article appeared in 1902:

> Looking from a certain hill top near Charlottesville, Virginia, one can see the birth-places of 3 men who gave to the United States nearly one-half of it's territory. Down to the East not far away was born Thomas Jefferson, whose Louisiana Purchase contributed to the union 13 states and territories. A few miles to the North was born George Rogers Clark who gave to Uncle Sam, Ohio, Indiana, Illinois, Wisconsin, Michigan, and one-half hours ride from Clark's birthplace was born Meriwether Lewis, who with William Clark, a brother of George Rogers Clark, and under the

At St. John's Church, in Richmond, Patrick Henry made his second famous appeal.

auspices of Thomas Jefferson, threaded an untrodden wilderness and plucked there from the Mighty Commonwealth of Washington and Oregon.[1]

There were those who paid a high price for independence and building a new nation of self government. The Clarks and Croghans were families of national heroes and the Revolutionary War had a great impact on their lives. For the Clarks, war was a mixture of tragedy as well as pride, fame and glory. Five Clark sons and five sons-in-law of John and Ann Clark were officers in the Revolutionary War. The third and fourth Clark sons, John and Richard, did not survive to celebrate the freedom they died for.

The eldest son Jonathan was a member of the famous Virginia Revolution Convention of 1775. As a delegate in the little church in Richmond, he heard the immortal words which rang out from Patrick Henry, "Give me liberty or give me death." Voices were raised against British tyranny.

After General George Rogers Clark's victory over Hamilton, he was given an ovation at the next meeting of the Continental Congress in Philadelphia and the feeling of all the patriots was expressed in the words of Benjamin Franklin, as he pushed his way through the congratulatory crowds, "Young man," he said, "you have given an empire to the republic."[2] Years later, the youngest brother William would proudly tell of this great moment in his brother's life.

The Croghan family came from Ireland in the 1740s and built an enormous empire of fur trade with friendly Indians and were responsible for opening up the Western frontier for the great tide of settlers coming from the East by building forts and settlements.

Lucy Clark Croghan's husband Major Croghan (pronounced Crawn) was no less a hero for the American cause of freedom. He had arrived from Ireland to America in his late teens under the protection of Sir William Johnson (a distant relative of the Croghans). Johnson was the most important British colonial leader of New York and Senior General of the English forces in North America. When the Revolutionary War started, young Croghan chose to join the forces of his new country, sacrificing opportunities he could have obtained through the influence of his British relatives.

Major Croghan served over seven years in the Revolutionary War and was with General George Washington's Army when the famous Delaware River crossing was made. He saw the worst side of war in snow, sleet, ice and hunger, but survived and became a wealthy Kentucky surveyor and frontier land owner on the Ohio River by accumulating property and wealth through land speculation and commercial trade on the Ohio and Mississippi Rivers. He established the town of Smithland, Kentucky at the mouth of Cumberland River and encouraged those from the East to build settlements in Western Kentucky.

Lucy Clark Croghan was one of ten children born to John and Ann Rogers Clark of Virginia. The father, John Clark III (1726–1799), a descendant of the first John Clark who came from England in 1620, was a planter on the James River in Virginia. The first John Clark provided and left a very comfortable lifestyle for three generations of Virginia-born Clarks.[3]

The first four Children of John and Ann Clark were born in Albemarle County, Virginia, near Monticello. The Clarks then built a home on land inherited from John's uncle in Caroline County Virginia. This new neighborhood was very expensive at the time.[4]

The Clarks of Virginia moved in the upper circles of society and were always welcome in the ballrooms of the "Genteel Tidewater Aristocracy". Their connections with the social elite were very extensive and they were associated through marriage or longtime friendships with Washingtons, Jeffersons, Madisons, Monroes, Lees, Hancocks, Randolphs, Harrisons, Taylors and Clarys.

These were men and families of education, culture and means who came to Kentucky. More than two hundred years later, descendants of the Clarks and Croghans are members of Colonial Dames, Society of Cincinnati and First Families of Virginia.

James Madison was a friend of Lucy's two older brothers, Jonathan and George Clark and all attended school taught by Donald Robinson who was married to the sister of their mother Ann Rogers Clark.[5]

The Clarks had many relatives and friends and visiting with each other was the essence of life. It has been said, these early Virginia planters "...lived to play and played to live-they will dance or die."[6] Daily events are recorded in Jonathan Clark's dairy, starting in the year 1770 while living in Virginia and ended 1811 after moving to Kentucky. High on the list of social activities while visiting in John and Ann Clark's Virginia home was dancing, although at other homes, some dances ended with a fight before daylight, as Jonathan occasionally recorded. The diary writings show an endless round of barbecues, fox chases, fishing, church meetings, and singings took place. But, most of all, the dances wherein queued planters' sons in satin coat and lace ruffles danced stately minuets and reels, they "trod a measure with the daughters of colonial dames."[7] The Clark home was the scene for many of these events.

Many years later, while living in St. Louis, Missouri, the youngest Clark son, Governor William Clark, never forgot the gracious living the Clark family left behind in Virginia and his formal gardens in Missouri were a fair duplication of the formal gardens and walks he had been accustomed to in his youth. A Clark nephew wrote, "We loved to hear of Uncle William's boyhood days in Virginia."[8]

Jonathan Clark's daughter Eleanor Ettings Clark Temple was born in Virginia in 1783. She remembered the old Clark house of her grandparents with beautiful grounds, ornamental trees and shrubs at the home-place in

Caroline County. Two special things stood out in her memory—the beautiful views from the home and the mammoth fireplace in the basement.[9]

Over one-hundred years had passed when a gentleman named Samuel J. Humphries purchased the old Clark place in Caroline County. The house was run down and very much in need of repairs. The landscaped gardens no longer existed and the grounds were overgrown. Humphries described the old frame house as being 42 feet long and 24 feet wide. The roof was 14 feet from the ground and there were five dormer windows, three on the south side of the house and two on the north. There were five fireplaces in the house, one in the cellar, three on the middle floor and one upstairs. The middle floor had three rooms, a hall and four closets. Two fireplaces on this floor were corner fireplaces. Upstairs were three rooms, two closets, one fireplace and a hall. All fireplaces fed off the large one measuring 44 feet in the cellar.[10]

When the Clarks lived there, the grounds at the old homestead had good soil for tobacco, corn, wheat, and oats. "The orchard was covered with varieties anywhere about, you could see it. The yard is called the prettiest any where about by all who see it on a level with a green sod and a beautiful shade of catalpa and mulberry trees."[11] One hundred years later, it was one of those houses that had the legend, "Washington Slept Here."

A family friend of the Clarks and Croghans, David Rozel Poignand, described the physical characteristics of some members of the Clark family which he had learned from John and Ann Clark's grandson Col. Charles Thruston, a nephew of General George. R. Clark. He wrote that George's mother had chastised George when he was sixteen—that she was a tall stout woman with red hair.[12] Joseph Rogers, a nephew of Ann Rogers Clark, wrote that his uncle John Clark "was a man of amiable excellent character, of sedate thoughtful appearance and not apt to say much in company."[13]

Jonathan was tall and well-proportioned; in his manners easy, uniform and engaging, and in his conversation, oftentimes sprightly, always agreeable.[14] Both General George Rogers Clark and Governor William Clark had red hair when young and both were about 6 feet tall.[15]

Ann Clark, who became Mrs. Owen Gwathmey was quite tall and had the same features as her brothers George and William.[16]

Lucy was small and petite and most resembled her brother Edmund. Today, one of her ball gowns is well preserved and is found in Colonial

Dames Dumbarton House, Washington, D.C., as part of their collection. The gown measures 52 inches from shoulder to bottom of hem and 26 inches around the waistline. It is found indexed under, "Clark, Lucy; sister of General George Rogers and William Clark." The name "Croghan" does not appear.[17]

Edmund Clark was a bachelor, was smaller than his brothers and had only a slight resemblance. He was much beloved until his death on March 11, 1815. The youngest in the family was Frances (Fanny), who was also like Edmund.[18]

Frances was the real beauty of the family and like William was spoiled by the older ones. She was widowed three times at a young age and the time spans between husbands proved to be very short. Her youngest brother, William, loved her dearly, looked out for her and gave special care.

The eldest Clark brother, Major General Jonathan Clark, served in the Virginia Militia and in the spring of 1781 was held a prisoner of the British along with Major William Croghan, a Revolutionary War officer, in Charleston, South Carolina. Major Croghan, who was on army parole, went to Virginia as guest of Jonathan Clark, who also was on parole. A relative wrote, "This was Major Croghan's first meeting of Lucy Clark when she was a fair young damsel just budding into womanhood."[19] Major Croghan spent several days with the Clarks in Caroline County, attending balls and parties. While there, he would also discover Lucy Clark was a very popular girl.

A nephew gave an account of Lucy Clark's young days while attending dances and socials. The following event was witnessed, retold and passed down by members of the Clark Family.

> Mrs. Croghan was very proud of her family, especially her brothers. On one occasion before her marriage, she visited the family of a Mrs. Holmes, who married a connection on the Rogers side." While there, she received such marked attention from the gallants of the neighborhood, that it excited the jealousy of the neighboring Belles. One of them spitefully said, 'She didn't see why it was that the men made such an ado over Lucy Clark, she was not overly handsome and beside had red hair.' This was reported to her, when she replied, 'I can tell her, they know my family and they know too, I have five brothers in the army, all officers, who have distinguished themselves in the service.[20]

Perhaps, a bubbly and vivacious personality was in Lucy's favor. Years later, her nephew Charles Anderson wrote, "I happen to know from John Symmes's letters that Benjamin Harrison IV (1755–1799) was courting Aunt Croghan and she rejected him."[21] Benjamin Harrison was eldest brother of President William Harrison and John Symmes was William Henry Harrison's father-in-law. In Jonathan Clark's diary, it shows Benjamin Harrison and Jonathan Clark were constant companions and visiting in each other's homes many times when young and carefree.

After Major Croghan's visit to the Clark home in Virginia, it is not known if he ever saw Lucy again until after the Clarks moved to Kentucky in 1784.

The war took its toll on the Clark family in Virginia. The third son John had been captured by the British and placed on a ship as a prisoner for the duration of the war. After the war ended, he was dying of starvation and disease, but was taken home to his parents in Virginia. George Rogers Clark was there for most of 1783 to be with his brother John in his last days. The parents, brothers, Jonathan, George, Edmund, William and the four Clark sisters stayed near their dying son and brother. Jonathan wrote in his diary something that isn't quite clear, although the last notation may have been in reference to a grave side funeral service.

> 1783—October 29, Rain—my brother John Died
> October 31, Clear—my brother John put in the ground at my fathers
> November 2, Clear—my brother John buried at my fathers

The following week after the burial, Jonathan left for Kentucky in 1783 with family servants for clearing land and cutting logs for the building of the two-story log house in preparation for their parents and family who would start for Kentucky in late 1784. He stayed near the Falls of the Ohio during this time, but soon returned to Virginia, leaving the builders behind to finish the house.[22]

George Rogers Clark wrote to Thomas Jefferson, February 8, 1784, "I shall set out for the Falls of Ohio in a few Days where I expect to reside perhaps for life."[23]

CHAPTER TWO

On the River to Kentucky

"1784 November 10, Accompanied my father & family on their way to Kentucky." Jonathan Clark.[1]

The Clark family was part of the migration pattern that was sweeping through the colonies. The Clarks were heavily influenced to move westward by George Rogers, who as a young man had done some exploring in Kentucky with two other friends. He was struck with the most beautiful and bountiful lands he had ever seen. The rivers, fertile soil, meadow lands for animal grazing, and the rocky cliffs and hills covered with forest captured his dreams and this is where he planned to settle. George Rogers wrote about his father, "I am convinced that if he once sees [the] country, he will not rest satisfied until he gets in it to live."[2] His father did return with George on his second trip and at that time he came to the same conclusion as George had about living in this "land of tomorrow."

After returning to Caroline County with his father, George again left for the frontier to make his home, but John Clark was concerned about his son living on the frontier where it was unsafe.

The word Kentucky, meaning "dark and bloody ground," was a paradise for the hardy pioneer who had heard the tales brought back from beyond "the blue wall," the Appalachian Mountains.

Once the decision was made to leave Virginia, the Clark family had a fixed destination clearly in mind before they left Caroline County, Virginia. Starting from Wheeling, West Virginia, they were going over four hundred miles away to hostile country at the "Falls of the Ohio." This land was considered part of Virginia.

> Owing to the badness of the roads, the inclemency of the weather, & the obstruction of the Monongahela with ice, did not arrive at the mouth of Kentucky until the 3rd of March 1785.[3]

The family was on the road and river for five months before they reached their destination. For those cold winter months, it would have been necessary to have a fireplace for warmth and cooking. The family, servants, as well as animals on the flatboat would have made an interesting combination of breathing creatures in very crowded quarters. One can only imagine how much food was loaded—staples, linens, feather beds, cooking utensils, spinning wheel and long-cherished heirlooms. Lucy's mother brought her silver teapot down the Ohio River, packed in a band (hat) box which now sits on the tea table in the parlor at Locust Grove. They would have had a hot cup of tea from time to time during those very cold icy days on the river.

Leaving Pittsburgh in a flatboat, John J. Audubon describes the scene while traveling down the Ohio River. The flatboats ranged from thirty to a hundred feet long and from ten to twenty feet wide. The boat had men, women and children, with horses, cattle, hogs and poultry. Vegetables and packages of seeds will fill space. The roof or deck of the boat was like a farmyard, being covered with hay, ploughs, carts, wagons and other tools for farming.[4]

Wheeling, Virginia was the place most flatboats were built and sometimes these were luxurious floating mansions. Some had dining rooms with fireplaces and fine paneling. Upon reaching their destination, these mantels and fine paneling would be stripped to be used in the new home. The flatboat would be pulled up on dry land, leveled on rocks for a foundation and the family would have lived in this until their house was completed. It is possible the building of the house on Beargrass Creek was completed when Lucy's parents, two young sisters, Elizabeth, Fanny and Brother William Clark (then 14 years old) arrived in Kentucky.

When the Clarks arrived at the Falls of the Ohio in 1785, they could have gone immediately to their new home or, after a long journey on the river, a short stay with others at Fort Nelson would have been more comfortable.

The distance from Fort Nelson to their new home was about five miles.[5] The trail or road went as directly as possible from the fort to Mulberry Hill, which would also be the first permanent home in Kentucky for General George Rogers Clark.

Many years later, Lucy's son-in-law George Hancock wrote about an event which took place soon after the Clark family's arrival in Kentucky. General George Clark, early one day, had sent a sergeant and ten or twelve men to guard the block house with a small cannon. Their orders were to keep out scouts and the moment any Indians or signs of their approach to fire the cannon as a signal. Lucy Croghan remembered the details of what happened when a signal came from the blockhouse which sent the people in and around Louisville, rushing from their fields to Fort Nelson located on the Ohio River. "General Clark sent men in every direction, and when nothing was discovered he sent an officer to know the cause. It was then ascertained from the sergeant of the guard that his wife had just been happily delivered of a child and he was celebrating the joyful event with a salute."[6]

Even though it was felt a time for celebration was in order since this was the first white child born in Jeffersonville, Indiana Clark could not brook such a disobedience of orders after producing so much alarm and trouble and the sergeant was dismissed from the service.

Mrs. Croghan added that she stood as godmother for the child and her faded memory seemed to think, it was Reverend General Muhlenberg who christened the child.

Soon after arrival in Kentucky, John and Ann Clark were visited with more sad news. Their son Richard had turned up missing on a trip to Vincennes sometime in February or March of 1785. His horse and saddlebags were found on the banks of the Wabash River. If Indians had taken him, they would not have left the horse and saddlebags. The family waited and waited, hoping he would appear one day, but this didn't happen.[7]

When the Clarks moved to Kentucky, they left behind the eldest Clark sister Ann, who was married in 1773 to Owen Gwathmey and living in Virginia with her family. John and Richard Clark had not survived the war and Edmund, a merchant in Richmond, Virginia, also stayed behind. Jonathan did not move his family to Kentucky until 1803. It was only George, Lucy, Elizabeth, Frances, and fourteen year old William who came to Kentucky in 1784.

CHAPTER THREE

Mulberry Hill and Frontier Life

" **M**ulberry Hill" would be the name of the Clark's new home in Kentucky. The home was a two story log house with a central hall on each floor separating four rooms on each floor. The kitchen was in a separate stone house—and a most important feature was a spring near by. The windows had heavy timber shutters which closed with bars. " ... The residence and kitchen were surrounded by a plank fence that also enclosed many large trees, reaching for less than half a mile from the residence to the public road was an avenue lined with locust trees on each side of a carriage way and foot path."[1]

The home was furnished with fine Virginia furniture. Before John and Ann Clark came to Kentucky, they sold their plantation in Caroline County and disposed of such of their personal effects as they felt it unwise to attempt to take with them. Among these was a beautiful Sheraton mahogany sideboard still preserved in the family of Ann's sister, Mrs. Lucy Rogers Redd, who purchased it.[2]

Some pieces such as the antique bedsteads and bureaus were brought through the wilderness after the house was completed.[3] Edmund Clark brought, or sent, some fine pieces of furniture, later by land. Merchandise was taken to Pittsburgh from Virginia by wagon, then, shipped down the Ohio River. However, there were cabinet makers in Kentucky at that time, whose pieces were of the same exquisite craftsmanship as those found on the Eastern seaboard.[4]

In 1799, William Clark inherited Mulberry Hill and all the furnishings. Before he left in 1803 for the Lewis and Clark Expedition, Jonathan Clark bought Mulberry Hill and furnishings for his son, Isaac. "Uncle Isaac," who never married, lived at the old Clark home-place until after the Civil War in 1868 when the slaves were freed. Temple Bodley, great-grandson of Jonathan Clark, described visiting his Uncle Isaac's house, Mulberry Hill.

In "Uncle Isaac's" time, the old two-story log house was famous in its day and always kept in immaculate order by his cook, old Aunt Rachel, and her son named Jake who served as houseboy. After every heavy meal, Jake had to rub the old mahogany inlaid dining table for what seemed to others to be a very long time since the legs and all were rubbed as well as the top.

This dining table belonged to Isaac's grandmother, Ann Rogers Clark. Isaac's parents were Jonathan and Sarah Hite Clark. Temple Bodley was writing a history and documenting the origin of family heirlooms for his two daughters.

> What has become of it [dining table] or of the other quaint old furniture he had, I don't know, except that I have his handsome old mahogany secretary [George Rogers Clark] which his father (and your great great grandfather) General Jonathan Clark had made by a French Cabinet maker from New Orleans in Middletown (12 miles East of Louisville on the Shelbyville Pike) in the last of the 18th Century when Middletown was larger than Louisville. It was secured after Uncle Isaac's death by Uncle John Pearce and I supposed was lost; but after the latter's death his son, George Pearce, while I was visiting him at Roxbury, the Spotsylvania Co. home of the Stanards afterwards bought by Uncle John [John Hite Clark] gave it to me. It was in the garret there and much battered, but I have had it fixed almost as good as new. Prize it. If you don't care for family relics of that kind, others of the kin, after you, may.
>
> There is a tiny silver spoon that I have too (marked J.C., I believe—I am away from home and in New Mexico at present writing). It was given to John Clark, your great, great, great, grandfather when he was a baby and kept by Uncle Isaac until in old age, he gave it to my mother and she to me. Also, there is an old silver small fat solid teapot marked, J.H.C. which, I think, belonged to my Mother's Uncle John Hite Clark and which I'll someday perhaps sketch on the opposite page. I jot these things down as they occur that you may know the history of these antiques when they come to your hands.[5]

George Rogers Clark Stuart, great-great-great-great, grandson of Jonathan Clark, was in possession of this desk and teapot when restoration of Locust Grove started in early 1960s. Stuart had wanted very much for the old desk and teapot to have a home where appreciation and special care would be given. These two pieces of family history were too precious to entrust to others, therefore, he traveled from Virginia with the "Ann Clark Teapot" and the Clark desk. Ann Clark's teapot sits on the tea table in the parlor at Locust Grove. The initials were engraved on the teapot by a family member to validate the line of descent—*J.H.C* (John Hite Clark, son of Jonathan).

The following letter (dated 1–31–99) was received from Rogers Stuart.

> Dear Mrs. Wilson,
>
> Responding to your request for information about the Clark teapot, there is nothing that I know which would constitute "hard" facts, except the date of the teapot, which as I recall is right for the time when it came into the family.
>
> Everything I know was learned from my mother, whose maiden name was Ellen Pearce Bodley, who died in 1969. Her father, Temple Bodley of Louisville, was a great historian, having written the *History of Kentucky and the Life Of George Rogers Clark*. He was a keen collector, particularly with respect to the Clark family, of which he was a member. At some point, Temple Bodley acquired the teapot, perhaps from his mother, Ellen Pearce Bodley, from Mulberry Hill, where some of the Clark family still lived. All I know is that my mother, Ellen Pearce Bodley Stuart, acquired the teapot either by gift or descent from my grandfather, and that she gave it to me. In turn, being a great admirer of Locust Grove, I gave the teapot to the Foundation.
>
> My mother never told me how my grandfather obtained the teapot. However, she did tell me several times that Ann Clark, wife of John, had brought the teapot to Louisville, in a band box, on a flat boat down the Ohio River. I assume that she obtained this information from my grandfather.
>
> Very truly yours,
> Rogers Stuart

A sound appreciation for history and ancestry is found in written material left by this family. As Bodley reminded his two young daughters, "... Isaac had kept the backgammon board that his father and uncle had used before this century was ushered in. In addition, he had preserved his mother's [grandmother's] bible, spectacles, toweling and bedding."[6]

For Lucy, those early years living at Mulberry Hill must have been happy ones with her two sisters, Elizabeth and Frances and two brothers George and William there with her. Some Virginia families living in Louisville between 1785 and 1789 were The Popes, Taylors, Bullitts, Harrods, Chenoweths, Bulgers, Pattons, Frenches, Brashears, Moores, Todds and others. These Kentucky families had sons and daughters of all ages.

Families on the frontier looked forward to any community activity which meant visiting and sharing any news from "home." In Virginia, the Clark sisters and William would have known all steps and movements of colonial assembly dancing. In 1786, Louisville, dance classes were being taught by Monsieur Nickle. Everyone practiced the *Paris*, which was leaping in circles, and the minuet with its graceful bowing and walking. In one dance, the dancers looked like strutting peacocks as in the minuet, the girls held their linsey dresses out like sails and slipped across the floor.[7]

On "election day," the men turned out at the designated voting place for barbecue, drinking and the latest news and gossip on the frontier. William Clark was old enough to be in attendance at these events. Having three sisters, William would have been included in schoolhouse barbecues and dances.

From time to time, William was sent on different missions against the Indians. At the age of nineteen, he was sent on his first military duty. This is when he served as Meriwether Lewis's captain and a lasting friendship developed between the two.

Between these short calls to duty, William would enjoy his days back home within a cheerful society where he had plenty of time to flirt with the young ladies of Louisville. In 1795, now Lieutenant William Clark was dispatched on a military mission on the Mississippi River. He wrote to Frances [Fanny], urging her to keep him informed about all the girls back home, especially, "Miss——." He claimed he suffered from long and painful absences from her conversation. William wanted to know all about the dances in Louisville and was very anxious to know about what girls were getting married.[8]

Christmas and other special days on the frontier were looked forward to by family and friends. Temple Bodley remembered the happy times at Mulberry Hill, especially, the "raggety breeches" made at Christmas.

Christmas in our household was a very happy time. A fruit cake
was being made and crullers (or "raggety breeches").[9]

Charles Anderson wrote about inflated hog bladders and how they
played an important part in celebrating Christmas when all the Clark
cousins were growing up in Kentucky. In the fall, the boys would sit on the
hog-pen fence while laying claim to the hog which would most likely have
the largest bladder.[10] A boy would blow very hard through a straw and the
bladder could become like a large balloon. When completely dry they were
painted with designs and hung on the outside of bedroom doors in the hall-
way. Early on Christmas morning, they gave off a loud bang when punc-
tured and simultaneously there was shouting, with a loud greeting of
"Christmas Cheer."[11]

A May festival enjoyed by children of English descent was celebrated
each spring in America for many years. There are cities in the south and on
the southern seaboard that celebrate maypole dancing even today. Pageants
and maypole dancing were very much celebrated in America up until World
War II.

It would be interesting to know, if in Virginia, the Clarks would have
observed the English tradition of gathering on the green for "May Day",
and whether they danced around the maypole. Nathaniel Hawthorne mentions "The Maypole of Merry Mount confirmed by Bradford."Washington
Irving also speaks of maypole dancing at that time.

William Clark's nephew wrote about early years (1840) in St. Louis,
"May Day seemed always to have been celebrated."[12] Reuben T. Durrett
wrote that "May flowers are still being strewn at General George Rogers
Clark's grave in [1893] at Cave Hill in Louisville."[13]

Today, garlands and wreaths of fresh flowers woven by tender
hands have been laid upon his grave, and his illustrious deeds
called back to memory.[14]

The youngest Clarks remained together at Mulberry Hill until Elizabeth
married Col. Richard Clough Anderson in 1787 and Lucy married Major
William Croghan in 1789.

George Rogers, William, and Frances remained with their parents at
Mulberry Hill. Frances was the pet of the family and continued to live with
her family even after her marriage. Her first two children were born at
Mulberry Hill.

George continued spending time in the woods in nature study. He wrote Jonathan that for several years he had lived almost retired, reading, hunting, fowling and corresponding with chosen friends while tending to his business. His sister Lucy, referring to this period, said her brother George "used to spend much time in reading, had a fine favorite horse he used to ride."[15] He had almost no contact with the outside world, except through correspondence.

Lucy was sixteen when Major Croghan first visited the Clark home in Virginia, but it would be eight years before she became his bride. No one knows how long she worked on her "hope chest" with William Croghan in mind, but a young girl would have started very early in making preparation for her future home by filling her hope chest with embroidered bed linens, window curtains, feather beds, and other household items. Fancy needlework would have gone into the sheets and pillowcases, as well as in the making of a beautiful museum-quality bride's quilt to be used on her "company" bed. The feather beds would keep her family from freezing to death during the winter. To get an idea of how much time was spent on sewing, one only has to read Abigail Adams's diaries and letters. She also, had sewn for five years before her marriage. The inventory in Abigail's hope chest was immense.

Major William Croghan and Lucy Clark were married the 14th of July, 1789, at Mulberry Hill. This was Lucy's special day and a time for celebrating, since these events had been very special in their Virginia home. After moving to Kentucky, the Clark family continued to gather for weddings, baptisms and other special occasions at John and Ann Clark's new home.

Jonathan's Diary shows that when his daughter Eleanor Elton was born and baptized (19 October 1783 in Virginia) at his father's and everyone was present, "...Mr. Gwathmey, my fathers & Brothers George Rogers & Edmund and my mothers & Loftiss[?] Gwathmey, Lucy & Elizabeth Sponsors."[16]

This christening event in Virginia was the last time the Clark parents and nine of their ten children would be together (Richard was not present). Ten days later brother John died.

When the new bride and groom William and Lucy married, there was a home waiting for them on Fifth and Main streets on the Louisville waterfront. In April 1790, William advertised the following in the *Kentucky Gazette* of Lexington, Kentucky:

> To be sold—a lot in Louisville, the best stand and situation for business in the whole town, with the following improvements; a two story log house 30 x 15 feet, well roofed and with an excellent ground cellar under the whole house; and another house one and a half story high, 24x15 with a complete store for goods at the end and a counting room or bed chamber at the other, extremely well finished, being circled all around with good plank well grooved and tongued, it has a complete loft and excellent stone chimney ... Cash or tobacco will be received in payment.[17]

CHAPTER FOUR

The Courtly Major of Locust Grove

The Croghans came from Wickland, Ireland, Isle of Man, which lies in the Irish Seas. The Dublin Directory of 1764 lists the Croghans as merchants and traders of Dublin having ties to the shipping business. They belonged to the Free-brothers of the Guild of Merchants.[1] The Major's father William, is listed under Commissioners of Affidavits, in Nov. 1763, on record in Dublin Offices.[2]

The world of trading and shipping in Major Croghan's background, served him well and to great advantage when building his commercial trade business on the Ohio River. The Croghan children heard their father reminisce many times about his homeland and family he left in Ireland.[3] His parents were William Croghan and Anne Heron and first daughter Anne Heron Croghan was named for his mother.

As a young boy growing up in Ireland, William listened to Irish relatives, traders and ship owners who had returned from that new far away land across the ocean. He must have dreamed of going to America as an apprentice for surveying with his famous Uncle George Croghan, especially, since "Croghan and his associates dreamed of towns and cities and commonwealths teeming with people rising where their chain carriers were running their lines."[4] A young astute William was aware of the importance placed on the western movement to the new lands.

Sometime before 1770, William left Dublin Ireland with relative Daniel Clark. He was sent to America by his father under the protection of Sir William Johnson who held the highest office in New York and also, great power among the Indian Tribes.[5]

Sir William Johnson had large landed properties and lived in baronial splendor at Johnson Hall. At first glance of a sketch of Johnson Hall, one is struck with the strong resemblance of Johnson Hall to present day Locust Grove in Louisville, Kentucky Sometime after this observation was made, it was discovered that William Clark's nephew, William Clark Kennerly, had made the following reference to Major William Croghan's country seat, "Locust Grove was similar to his ancestral home in Ireland."[6]

At Johnson Hall, Sir William had wide interests and there were fine pictures, the best books on history, literature, philosophy, and science in the library. Scientific experiments took place in one room and a correspondence continued with Benjamin Franklin. He was one of those who promoted arts in America and helped to launch Rutgers and Columbia Universities. Sir William made arrangements for the release of deserving men from debtors' prison, giving them a fresh start.[7]

Shades of Sir William Johnson's character seemed to be born and bred in Major William Croghan's makeup. He too, would reach out to others when an undeserved misfortune came their way.

There is evidence which suggests that when William Croghan arrived in America "under the protection of Sir William Johnson", he could very well have been in the training program at Johnson Hall where Sir Johnson prepared young men for an 'arduous career for an active life';

1st. You will keep your Party sober and in good order ...

2nd. If any difference should arise ... I am to be immediately acquainted with it.

3rd. The Sergeant to take care that the Men's Quarters be kept very Clean and that they wash well, freshen their Salt Provisions, the neglect of which makes them subject to many Disorders.

There were other instructions such as: keeping a sentry at each corner of his house and others at the gate of the fort on the outside.[8]

The famous Uncle Colonel George Croghan, (1715–1782) was one of the legendary figures of the Ohio country. He was a true Irishman and one of the greatest of the Pennsylvania traders. The Colonel was with George Washington in the battle of Fort Necessity. As deputy superintendent of Indian Affairs, with headquarters at Fort Pitt, he bought furs, sold goods, and made and lost money. He also made friends with the Indian tribes and built a wilderness empire. Sir William Johnson could not have gained so much success with the Native Americans, had it not been for Col. Croghan's friendship with the Indians.

At that time in history, the Ohio and Mississippi were great arteries of commerce for English forces of expansion, but their main purpose was for trade with the Indians. England and France were competing to control that trade on the frontier. After the end of the French and Indian War in 1763, England gained sole possession of the northwest frontier.

After the British had permanent possession of the vast western territory, Colonel George Croghan was called upon to open the West to English enterprise. In 1770, Uncle George Croghan opened a land office at Pittsburgh and began selling his lands to anyone who would buy.[9] [Twenty years later, Major Croghan would do the same at Locust Grove in Kentucky.]

Croghan realized help would be needed to set up army posts and settlements, but he had one great advantage—he knew better than anyone the surrounding region and its natives. Col. Croghan would have to find a staff of assistant agents including interpreters and gunsmiths, several clerks, and finally, a surgeon for each post. Also, he needed to secure land grants to erect English forts; gardens and cornfields had to be planted, and pastures for the needs of the garrison were necessary.[10]

Irish relatives heard "the call" for assistance in taking on the enormous operations needed for settling the new vast territory, and came to America for land speculation. Most important, they all had a common interest in developing the great wilderness and George was their leader.

"He was an attractive man, hot-tempered, a lover of wine and women." This statement was written about Col. Croghan by Pittsburgh's first diarist, James Kenny. Also, written on 1759, August 20th, "Croghan has a black eye this morning, and I have been informed, that he was drunk and fought with ye Indians, and the Teedyuscung gave him ye black eye."[11]

Uncle George Croghan's dwelling, "Croghan Hall" at the foot of Lake Otsego, [New York] was built by indentured servants, a mason and skilled carpenters. The log building had figured wallpaper [put up with paste and flour], his table was covered with the best damask cloths, ivory-handled knives and forks were at each place and thirty chairs for dining table. The six fireplaces had their sets of andirons, shovels, tongs and glass for the windows; brass locks for the doors. Col. Croghan had planned to retire as a country gentleman, but this didn't happen.[12]

Young William Croghan was ambitious when he arrived in America and his dream of becoming a land owner must have been uppermost in his mind. One way to acquire money for purchasing land was to join the British army. Uncle George Croghan's daughter married a British officer, Lieutenant Augustine Prevost, son of the British general, and he could have encouraged William to join the British military on February 22, 1771, where he served until late in 1774.

In 1775, William Croghan bought 6,424 acres of land in Washington County, Pennsylvania, paying his Uncle George Croghan, 702 pounds sterling.[13] His army pay must have been saved for this purchase and it is possible he spent the following months working for his Irish relatives as chain carrier and at the same time, learning the basics of surveying. But then, William could have become a clerk to Uncle George as did the Croghan kinsman Daniel Clark, who became the most prominent American in New Orleans.[14]

The Gratz brothers who were traders and merchants in Philadelphia were also European traders before coming to America and doing business with Uncle George Croghan.

> The Gratz brothers were particularly associated with William Croghan's relatives, Uncle George Croghan, the Indian trader and another relative, William Trent. Michael Gratz came to the United States in 1755. His brother Barnard had arrived in this country one year earlier. The Gratz brothers played a large role part in building commercial activities of Philadelphia and the country. Their specialty was the fur trade connected with William's Uncle "Colonel" Croghan.
>
> The Gratz brothers' trading business was formed with merchants and traders in Europe, but they were two of the first signers to protest against Britains Stamp Act and wouldn't accept any goods shipped from Britain.[15]

William was closely associated with the Gratz brothers when the British passed the Stamp Act. The brothers reacted to this by declaring their loyalty to the new country and by joining and signing a protest circulating by the revolutionaries in their cause against the British. William also, made a decision to fight for a country which had been good to him. He joined the American army in 1775 and on April 19, 1776, was commissioned a captain in the 8th Virginia regiment.

The next two years, Captain William Croghan survived exhausting difficult conditions while fighting the British. During the war, William kept a diary which reveals the horrors of war, deprivation, sickness and torture by Indians if captured.

On Christmas Eve, 1776, Croghan and his men crossed the Delaware River with General Washington. They overcame the Hessians, taking over 700 prisoners during the worst day of sleet and rain that could be. During the next few days the revolutionaries forced the British soldiers to retreat to Trenton, marching through snow, hail and rain, again in very severe weather. Captain Croghan wrote about the many soldiers who were without shoes and were wearing summer clothes.

Under the leadership of General Nathaniel Greene, Captain Croghan and his men continued to be involved in battles. On the 1st of February, 1777, the British beat the Americans soundly. Food, clothes and supplies were desperately needed and the number of desertions was great. The war went on and at times, it "was barbarity to the utmost."[16]

After surviving a dreadful winter at Valley Forge, and always in need of clothes, food and materials, Washington's army was desperate. It wasn't until spring that the rag tag army received supplies and went through rigorous training under Captain Baron von Steuben.

There were bright moments of relief as Croghan and comrades marched through New Jersey and Virginia. They were treated well and requested to fire their cannons for the citizens as they passed through. At Brunswick "There will be Several great Entertainments, Balls, Concerts, &c, &c. You see we Can find Methods to Spend Money in Camps, & I think my proportion will be no Small sum for two Weeks to Come."[17]

At Fredericksburg, Croghan stayed in taverns or private homes most of the time. The troops remained ten days, where "we spent in the Most Agreeable Manner the Inhabitants doing all in their power to Add to our happiness; we had Several public & Private Balls, a Constant Round of Invitations on hand & found a hearty Welcome to All houses." On February 11th, Washington's family had the soldiers to join them in Washington's Birthday celebration, including a dinner and large ball. When they reached Richmond, the troops celebrated again, with concerts, public and private balls.[18]

In May 1780, the situation had changed for the Southern Army which meant surrendering peacefully to the British and becoming prisoners of war.

Providential favor came William Croghan's way when British officer Major Augustine Prevost, son-in-law of Col. George Croghan, saw the name "Croghan" on the list of prisoners, recognized William as his father-in-law's nephew and sent for him. At Major Croghan's request, Prevost gave permission for Jonathan Clark, his Colonel, to accompany him on parole of honor which meant, while both were prisoners of war they were granted permission to travel within the limits of a certain area.[19]

Major Croghan shared the good news with his Uncle George Croghan in the following letter, even though, it would be February 1781 before he and Jonathan would be allowed to visit Jonathan's family.

Charlestown South Carolina, June 10th, 1780

Dr. Colonel

Major Provost Setout from this place from Savannah yesterday, he came here from georgia a few days ago with his Father who has Imbark'd for England—the Major is to be Stationed here & has not gone for Mrs. Provost & Expects to be here with her about two Weeks hence—I am on Parole at a place Called Haddrells Point Near this Town, but got permission to Come See the Major, who has been Exceeding kind in Endeavoring to get me a parole to go to Northward & in Every Respect to Serve me,—with his Interest I hope for the pleasure of Seeing you in two Months. Mrs. Provost is well—but has no Children Living but George & Augustin, who are in England & Officers in the Regt. with the Major—I am Just going to Cross the River to Join the Other Continental officers Prisoners at Haddrells point.

(Add on at bottom of letter)

Savannah. 26,

Shall do Myself the pleasure of Writing you Every Opportunity & am With Due Respect. Your Much obliged & Most Humble Serv. W. Croghan

Shortly After my Arrival here I sent you a Letter from the Major [Provost] which I Received from an Exchanged Officer who Came here.[20]

One of those at Yorktown when Cornwallis was hopelessly trapped by the Americans was Major Croghan who witnessed the surrender of the British on October 19, 1781. The British band played "The World Turned

Upside Down." The redcoats hated the despised rebels and wanted to surrender to the French, whom they considered gentlemen. Washington said "No" and to add to their humiliation, Lafayette ordered the band to play "Yankee Doodle".[21]

Major Croghan's presence at Yorktown entitled him as a charter member (1783) of the Society of Cincinnati, a fraternal and patriotic organization. Major's portrait proudly shows the ribbon across his chest bearing the insignia of Society of the Cincinnati.

All the fighting on American soil ceased, but the treaty of peace was deadlocked for three years and was not signed until 1783. The American Commissioners were holding out against Britain who was demanding all confiscated land taken from the Tories to be returned and all territory between the Ohio River and the Canadian Border returned and the Canadian Border moved back to the Ohio River. This meant all the Northwest Territory won by General George Rogers Clark would be returned to the British.

Benjamin Franklin and John Adams didn't place too much importance on that area, but John Jay held out to the end. He negotiated and the Commissioners had to agree to the final treaty giving America claim to the Northwest and extended the boundaries of the young nation to include all the land between the Appalachian Mountains and the Mississippi River, from the Great Lakes south to Florida. For this, Commissioners had to restore all land confiscated from Tories and Loyalists.[22]

Before the treaty was signed, the returning Revolutionary War veterans had immediately began laying out the river front in Louisville, and included a park with some original trees before selling the lots.[23]

In May, 1786, the Louisville Trustees were forced to sell the public lands into lots for John Campbell who was pressing without mercy for payment to the Tory John Campbell. After recovering the pay, he brought in another claim which the Legislature of Virginia allowed him to collect from the sale of Lots.[24]

Col. George Croghan and associates Barnard Gratz had petitioned Virginia to release their lands which they had bought and paid for, but to no avail. In 1779, other land owners were denied patents from Williamsburg, only because they were in the service of their country, preventing them from making such improvements as would have secured for them a quantity of land.

Petition of Major William Croghan [1784]
To the Honorable Speaker of General Assembly of Virginia

Humbly [illegible] that hearing the opinion of Several able
Lawyers of this State on Colonel George Croghans title to lands
he claimed in the neighborhood of Pittsburg and they alleging it
to be good, Your petitioner on Viewing part of said lands on
Raccoon Creek and River Ohio amounting to 6420 Acres
Unsettled and claimed by no Other person, did in February 1887.
purchase it from him giving to the Amount of (illegible) Sterling
for it, and was in actual possession there of, and about Improving
it,—but having Inlisted a Company of Men and with them being
Ordered to Join the army in which he has continued to the 3rd
day of this Month. This being the Case your Petitioner could not
attend to his land by Improving it or Other ways looking After it,
by which Means Others have Actually Settle thereon and obtained
titles for it, to the Manifest Injury of your petitioner.[25]

Major Croghan had to accept the fact that reclaiming his land was a lost
cause, but received $7,149 for his military service when released. It's easy
to assume that the Major had plans, knew what he wanted and how to
achieve his dream. First in order was a trip to Williamsburg where he
passed an examination and received his certificate as a surveyor from the
College of William and Mary. William also delivered George Rogers Clark's
surveyor's commission to him.

Coming from a background of merchants, trading and shipping, William
Croghan was a man who would have had all future plans laid out carefully and commercial trade on the river would be uppermost in his mind.
River avenues and the difficulties of land communication tended to keep
commercial operation close to the plantation houses. He built his wharf on
the river and owned a ferry boat for crossing over to the Indiana side and
had other boats for shipping his produce down the river to New Orleans.

William Croghan's interest also, lay in land speculation and a vision of
developing the lands for settlements. "Colonel George Croghan's dreams
helped to inspire such younger men as Daniel Clark, John Campbell,
Dorsey Pentecost, William Thompson, Barnard and Michael Gratz, and
William Croghan to carry on similar work on more distant frontiers."[26]

In August 1781, while at Fort Pitt, the Major was already showing interest in the building of settlements in the Ohio Valley territory and was at a loss to understand why General Clark was unable to raise the 1,500 men needed for driving the Indian tribes from Kentucky, which was only a county of Virginia at the time. "I have no doubt the people on this side the mountain in particular would be sencible of the advantage they must reap, by being Able to live at their plantations without the dread of being Scalpt....The reason so few went with him from this plac, is owing to the dispute that subsists here between the Virginians & Pennsylvanians respecting the true bounds of the latter."[27]

Later, Major Croghan spent years in Western Kentucky surveying the lands appropriated to Revolutionary War veterans of the Virginia state line which finally took its toll health-wise, but Croghan became a wealthy man. Surveyors were called locators because they located the surveys at their own risk when entering lands where white men had never been before. Richard Anderson's son Charles wrote about how these locators became enormously rich. At first the locator's fee was one-half and afterwards one-third of the individual tracts surveyed.[28] At the same time, there was great danger of someone finding their scalps suspended to their compass. The following has been written about Major Croghan.

> The town of Smithland had been planned and established as a profit-making venture by William Croghan, Sr. of Locust Grove, near Louisville. ... Croghan was a gentleman of widespread influence in Kentucky. ... Most of the boats descending to New Orleans and Memphis generally make a halt here, either for hams, provisions, boats or repairs. ... there were two ferries at Smithland, one on the Ohio and the other crossing the Cumberland.[29]

Major Croghan was the one who influenced the migration from Virginia and South Carolina to Western Kentucky, especially those who settled in Muhlenberg, Hopkins, Caldwell and Christian Counties along Green River and Trade Water River. In 1808, the Wilsons, Clarks, Foxes, Sisks, Kirkwoods, Gordons and many others settled in Hopkins County and their descendants live there today. John Wilson built the old landmark "Wilson's Warehouse" on Tradewater River in Hopkins County. Boats would come from other waterways down Tradewater River with tobacco. Upon arrival at Wilson's Warehouse the tobacco would be stored in the ware-

house until a larger boat would arrive for shipment down the Mississippi to larger markets.

In 1806, Thomas Jefferson's sister Lucy Jefferson Lewis and family built their home in Smithland, which was established by Major Croghan at the mouth of the Cumberland.[30] One of President Jefferson's unfilled dreams was to visit "...the lands of Cumberland. I do not set it down as impossible that I should see you there. I have never ceased to wish to descend the Ohio and Mississippi ...."[31]

CHAPTER FIVE

Locust Grove

Between 1775 and 1800 many Kentuckians were killed by Indians. One of the worst Indian massacres took place in Laurel County in 1786, just one year after the Clarks moved to Kentucky. Some thirty men and women were killed. The Indian massacre of the Chenoweth family took place in the summer of 1789 near Middletown, a few miles from Mulberry Hill, an event that took place four days after Lucy's wedding. Lucy's brother-in-law, Col. Anderson, and her young brother William were in the rescue party.[1]

Regardless of the dangers, sometime during the first two years of marriage, William and Lucy moved from the house downtown to the property overlooking the Ohio River. They lived in a cabin on the lawn while the house was being built. This would be a country estate called Locust Grove.

While living in this cabin, the first Croghan child John was born in April 1790. The next year, Lucy had quite a scare, while the Major was away. She was left alone with a few servants and a baby son. Some of the following is not quite clear, but is a direct quote as written;

> Mrs. Croghan, when she was left alone at her home (or farm) 5 miles above Lou, she went outside at dusk one evening to bring in some cloths from a line or bushes, fearing the Indians might steal them. While out there, she discovered an Indian (without herself having been seen by him) creeping along on the ground from the stable towards the house. She could not run to the house without being seen, so she dodged into a clump of bushes. The Indian passed within six feet of her and went around the house. Mrs. Croghan had left the door front open & fire light shined out and gave light. She could see from her hiding place in front open and through the door, that the Indian opened the wide window blind in the small rear window

in which was no sash or glass and make discoveries—and then darted off out of sight when Mrs. Croghan, dodged into the house and fastened the door and windows, blew the horn, which was a signal of alarm and shortly several Negroes on the place hastened there. The Indian had evidently reconsidered and hearing the horn and Negroes coming—saw they were alarmed and prepared—so went off without attempting any mischief.[2]

On April 26, 1791, the Major placed a notice in *The Kentucky Gazette*.

Notice is hereby given, that agreeable to direction from the superintendents of the Virginia State line, I will open the office at my house near the falls of Ohio, on Monday the 1st day of August next, to receive entries, and proceed to surveying the military state lands on the south side of Green river &c.[3]

Other exciting and interesting events took place during this time period. In 1791, a very young Zachary Taylor was living with his family, the Richard Taylors, on adjoining property to Locust Grove. Their house was in sight, but yet, quite a distance away. Lucy was visiting the Taylors when an Indian alarm occurred causing those at the Taylor home to determine it was safer for her to remain with them. She was confined there and gave birth to her son, George Croghan. It was some time before she could return home to safety.[4]

This incident was told by Lucy's son-in-law, George Hancock to Lyman Draper in 1868. There were family connections between Hancocks, Taylors, Clarks, and Croghans. William Clark married George Hancock's sister, Julia. Zachary Taylor and George Hancock's mothers were sisters. Hancock would have heard such family stories from more than one source.

The Croghans' country seat Locust Grove was one of the finest in Louisville. Unlike the southern planter's estate whose riches came from a single crop such as tobacco or cotton on several hundred acres, Locust Grove was a self-sustaining village cut from the forest. This economic unit sold produce from orchards, hams, diary products, and tobacco, which was loaded on the wharf and sent down the river to New Orleans. In the center looking out upon broad acres, stood the mansion of Locust Grove, a large brick home of the Georgian style resembling those of the wealthy country gentlemen of England. In 1792 Major Croghan was being taxed for 516 acres, 1 white male, 17 blacks, and 34 horses and cattle.[5]

Major William Croghan, 1822. Painted by John Wesley Jarvis.
(From the collection of Historic Locust Grove, Inc.)

Lucy Clark Croghan, 1822. Painted by John Wesley Jarvis.
(From the collection of Historic Locust Grove, Inc.)

Much was required to organize and administer an enterprise of this kind, so heavy responsibility fell upon the planter and his wife. However, very soon, these former Virginians would be living a luxurious life on the Kentucky frontier and their homes would be filled with table silver, furniture, clothes, wines and books. Wallpaper ordered from France was used in the ballroom at Locust Grove.

Lucy's sister, Elizabeth and husband Richard C. Anderson lived a few miles south, in their country seat "Soldier's Retreat." The sisters had married business partners who were away from home for long periods of time, although if emergencies arose, other family members lived in close proximity and could be sent for easily. In the meantime, life at Mulberry Hill was still very much alive.

The young beauty, Frances Clark, was in love with a captain from Virginia whose family was well established in the tidewater aristocracy. This captain, Charles Thruston, had left Frances in Kentucky with a promise of returning for marriage. She waited and waited and heard nothing from him, which convinced her that Thruston was no longer in love with her and did not wish to marry. Brokenhearted, on February 21, 1791, Frances married Dr. James O'Fallon (1749–1793) from Atholone, Ireland. At the time of marriage, Frances was seventeen years old. Her new husband was twenty-four years older.

Dr. O'Fallon emigrated from Ireland to America and landed in Charleston by way of the West Indies in 1775. "By his 'writing', incited the people to arms for which was imprisoned by the British Gov. ... soon liberated by 1800. He was a most Skilful Physician & Surgeon and was put at the head of the Hospital dept. with pay of Colonel, which he held to the end of the war."[6] The first O'Fallon son John was born at Mulberry Hill, November 17, 1791. The second son Benjamin was born September 8, 1793, one month after his father, James O'Fallon had died in August. Frances, a young widow with two small boys, continued to live with her family at Mulberry Hill.

This marriage was not a happy one. O'Fallon had been so abusive to Frances, her father John Clark had sent him away from his home in the middle of the night.[7] O'Fallon was living alone in Louisville when he died two months before the second son, Benjamin was born. His meager estate was auctioned and sold to several different buyers. There were some surgical instruments, personal items, and many housekeeping items for sale.

Major William Croghan was present and purchased some small pieces which had belonged to his old friend. Since there was no next of kin other than Frances and infant sons, Major Croghan retrieved and saved some items which he passed on to John O'Fallon when he became twenty-one. Included were papers with information about O'Fallon and Croghan experiences when coming to America.[8]

Two years later, sadness struck John and Ann Clark when their daughter Elizabeth died. The following is taken from *The Diary and Journal of Richard Clough Anderson Jr.,* son of Col. Anderson of Soldiers Retreat.

> It must have been in 1794, in the fall of that year, my father went in company with Major William Croghan & to Virginia & returned on the night of the 8th of December. My sister, Betsey (Elizabeth) was born on the 7th, the day before. My Mother was never well afterwards, she died on Thursday 15, January 1795, aged 27 and some months. I had been taken to my Grandfather's several days before, but returned with Mrs. James O'Fallon on the day of her death.[9]

Frances was a young widow when Elizabeth died and William had been dispatched to Mississippi before his sister's death. Following the death of his mother, Richard Clough Anderson, Jr. was sent to live with his grandparents John and Ann Clark at Mulberry Hill. While there, he joined the O'Fallon cousins, receiving his schooling by tutors.[10]

He lived with his Clark grandparents for almost two years before he returned to his father [Richard Clough Anderson] at Soldiers Retreat and his father's new wife. A schoolmaster was hired for Richard's education until 1800 and then Richard was sent to Virginia to finish his schooling.[11]

When Frances married O'Fallon in 1791, Charles Mynn Thruston, her first love, was devastated when news of Frances' marriage reached him. For some reason, he was preoccupied in some endeavor back in Virginia which made it difficult to correspond with her. After the death of Dr. O'Fallon, Frances Clark O'Fallon was free once again to marry Captain Thruston.

Frances married Thruston in January 1796, one year after her sister Elizabeth's death. The O'Fallon boys were left to live with their grandparents and uncles at Mulberry Hill while Frances and her new husband went to their new home, located a few miles from Louisville at a place called Westport, overlooking the Ohio River. The O'Fallon boys were four and three years old at that time.

Col. John J. O'Fallon (From the collection of the Missouri Historical Society. Painting by George Calder Eichbaum)

Major Benjamin O'Fallon (From the collection of the Missouri Historical Society. Painting by Astley Cooper)

Charles William Thruston (courtesy of Charles Timothy Todhunter)

Frances Ann Thruston Ballard (courtesy of Charles Timothy Todhunter)

After four years of marriage, and two more children, Frances' husband Captain Thruston was killed on December 11, 1800, by his body servant Luke, who feared his master would punish him for repeated misdemeanors. Capt. Thruston had made a trip to Virginia, not allowing Luke to go with him and leaving him with a warning that there would be a good thrashing upon his return if any misconduct occurred during his absence. Knowing that he had not attended to his duties, Luke ran away, but on Thruston's return in December, a servant reported that Luke had been in the kitchen and had stolen a leg of lamb. Capt. Thruston and his small son went out looking and found Luke's footprints in the snow. Luke was hiding in a corn shock and upon discovery, sprang on his master, stabbing him in the back with a carving knife, which he had stolen from the kitchen.[12] Needless to say, Luke was caught and hung. Again, Frances was a widow, with two more children.

This was a difficult time for Frances. Her mother Ann Clark had died two years before on December 24, 1798 and six months later, her father John Clark passed away on July 29, 1799. Both were buried in the small family cemetery at Mulberry Hill.

When Ann and John Clark died, Major William Croghan made his heart and home big enough for the two sons of his old friend James O'Fallon and he brought the O'Fallon boys, age eight and seven years old, to Locust Grove.

After Thruston's death and before Frances married Dennis Fitzhugh in 1805, she moved around and lived with other family members. John O'Fallon wrote about time spent at their Uncle George Clark's cabin in Clarksville. "I recollect being present at Mulberry Hill and afterwards at Point of Rocks, when visited by Indians, chiefs & Braves. A celebrated chief called [unintelligible] was the principal one, but I was too young to comprehend anything on the subject."[13]

Even though they may have spent some time with their mother while visiting relatives, after the death of their Grandparents, John O'Fallon would always refer to himself and his brother Benjamin as orphan boys. "Being an orphan was sent away and joining the army, seldom afterwards with my relatives."[14]

Since their mother and Thruston did not include the O'Fallon boys in their household at Westport, Kentucky, their Grandfather Clark knew the two boys would need help and included them in his will:

William and grandsons John and Benjamin O'Fallon, to be equally divided, 3000 Acres 'which I claim under an entry on Treasury
Warrant' 7926 made in surveyors office of Fayette County, March
29, 1783, surveyed and patented in his name." To grandsons,
John and Benjamin O'Fallon when they become of age, 100 and
50 pounds, respectively, also certain Negroes.[15]

The youngest Clark son, William, inherited Mulberry Hill with all its
furnishings and some slaves. William was also, appointed guardian for
administering the O'Fallon boys' inheritance from Grandfather Clark.
Some arrangements or an agreement had been made with William to
finance their education while they attended school with their cousins in
Danville, Kentucky. William's approval was needed for any expenditure
such as their school supplies and clothing.

At that time, a new wife and an appointment to The Bureau of Indian
Affairs kept William preoccupied in St. Louis, and he neglected the
nephews' needs. Letters sent to him were not answered, so John O'Fallon
had to appeal to his Uncle Jonathan and new stepfather Dennis Fitzhugh
for the money which was meant for his brother's and his education. When
William Clark replied to Jonathan and Fitzhugh's request for reimburse-
ments, his reply was always that he had sent the money which was lost in
mail or he had written a bank draft on Jonathan or Fitzhugh until he could
cover the amount.[16]

John O'Fallon, unlike his cousins, was without a horse at school, but
Uncle George Rogers Clark didn't forget his nephews. John and his uncle
exchanged letters and in one (November 18, 1808) John thanked his uncle
for the horse he had sent.[17]

Many years later, in 1849 when John O'Fallon had become one of the
wealthiest men in St. Louis, the bitter memory of being a helpless "orphan"
remained with him.[18]

When Thruston died, William was also made administrator of the
Thruston estate left for Frances and the two Thruston children. Again, due
to William's rationale that he deserved a good size commission for administering the estate, he was taking "his" share of the money. Frances sent a
letter to William informing him,

> I never thought to mention this morning who I wished to be my
> Executor. I wish you to name John [O'Fallon] if you think of
> another nephew put him down. I thought I had so little property

that one would be sufficient. It is my wish that the court would appoint my son John guardian for my two youngest children. Your Sis.[19]

Again it became necessary for Jonathan to intervene on behalf of his Thruston niece and nephew in receiving their inheritance. Because of this, there were some strained feelings between Frances and William for a short period of time.

Two years after William Clark's departure on the famous Lewis and Clark Expedition, Frances married Judge Dennis Fitzhugh on May 13, 1805. Judge Fitzhugh was a merchant and trader with his place of business on Fifth and Main in Louisville and living quarters above the store. Two more children were born to the Fitzhugh marriage, bringing the total of Frances's children to six.

Locust Grove at that time was a center of energetic activity. Lucy was twenty-five years of age when her first child was born and forty years old when the last one arrived. Much of her time was spent in the nursery and caring for the sick, especially in the servants' quarters. Keeping the servants alive and well was important, since this meant financial gain for her husband.

Locust Grove was open for visitors at all times. The following is found in *Sketches of Christ Church Cathedral,* Rector James Craile.

> Major Croghan who emigrated from Vir to Ky. 1782. Then and for a long time after Major Croghan was a Registrar of the Land-Office, and the primative building in which he kept the records and discharged the function of his office, was still preserved with reverent care in the centre of the garden at L.G. the family seat, seven miles above the city. Major Croghan and his family were celebrated far and near for their unbound hospitality. The house was the home of every stranger who visited Lou. and the constant resort of the neighboring families of town & country.[20]

Virginia relatives, coming to Kentucky, were especially welcome and Locust Grove would be their first stop on the river. They were always welcome at their Cousin Lucy Clark Croghan's house and her husband Major Croghan, a country gentleman, also extended a gracious welcome.

From Goochland, Virginia, came Joseph Rogers Underwood, a boy of twelve, who would later be a member of the U.S. Congress, both in the

House and the Senate. Joseph wrote that when he was a young boy, his Uncle Edmund Rogers had made preparations for a trip to Kentucky. With a wagon and team, a riding horse, and some half-dozen slaves they passed through Fredericksburg, which was the first town he had ever seen. At Brownsville on the Monongahela, his uncle purchased a boat for floating down the river to Louisville. Before reaching Louisville, he was "… put on shore about six miles above Louisville on the left bank of the river, with directions how to find the residence of Major William Croghan, after which my uncle passed on with the boats to Louisville."[21]

Joseph found Major Croghan at home and introduced himself as the relative of his wife and children. He was received with open arms. The Major ordered his horse and went directly to meet Joseph's Uncle Edmund in Louisville. Joseph remained for some days at the Major's home while his uncle made arrangements to resume the journey. He was sent for, passed through Louisville, then a mere village, and journeyed on, settling on Barren River near Edmonton, in May of 1803.

Lucy and family must have been elated upon seeing a cousin from Virginia. He would have been plied with questions about friends and relatives left behind. Visiting filled the needs for stimulation on the frontier. Family ties were strong, and there was much open expression of affection. If you were four times removed, you were still addressed as "cousin." Even in the 20th century, southerners referred to their kin as "kissing cousins."

Heavy burdens and responsibilities did not take precedence over the welfare of the Croghan children at Locust Grove. Major Croghan took great interest in his children's education. He wanted the best for them and his letters were full of encouragement and explanations why their education was so important. The three oldest sons were sent to the finest schools in the country.

Education began in the ballroom at Locust Grove with daily recitations and music lessons under the guidance of an instructor who sometimes lived with the family. Before the Clark, Croghan, Anderson and O'Fallon cousins went away to school, they probably came together "… at the schoolhouse near Colonel Taylor's." This schoolhouse was built before 1808 and was within riding distance for neighborhood families.[22]

The boys bowed to the schoolmaster when entering the room and the girls dropped courtesies. The students studied aloud and were audible at some distance. Men who began their education this way defend the vocalizing, saying that it forced concentration and strengthened the voice.

Students sometimes sang their lessons which were set to music of old English tunes. In 1806–1807, John and George Croghan and the O'Fallon, Gwathmey, and Anderson cousins attended the Danville Kentucky Seminary.

Both Croghan sons, John and George, attended the College of William and Mary, graduating in 1810. John Croghan went on to study medicine at the University of Pennsylvania. His M.D. degree was awarded in 1813.[23] George Croghan graduated from the College of William and Mary in 1810 and returned to Locust Grove. He studied law for a while, but soon realized it wasn't for him. He was well read, but growing up listening to the family stories of war battles, he had the military life in his blood. Somewhere, out there, greater and more exciting things were waiting for him. The timing was right and there was opportunity for George Croghan to prove his greatness in war.

The Kentucky governor called for volunteers to fight the British and Indians around the Great Lakes. General George Rogers Clark wrote that his nephews George Croghan, John Gwathmey, and John O'Fallon volunteered for this mission and together they left Louisville in the fall of 1811 for the Great Lakes area.[24]

William Croghan, Jr. attended Transylvania College, Lexington, Kentucky, graduating in 1810. He then attended Dickinson College, in Carlisle, Pennsylvania from 1811–1814, studying scientific training, chemistry and philosophy, and became a member of Union Philosophical Society. He also attended Litchfield Law School at Litchfield, Connecticut in 1815 and 1816.[25]

Major Croghan wasn't well during this time and after graduation, William Jr. came home to manage Locust Grove and was involved in many other family interests, such as the Bank of The United States. He ultimately inherited Locust Grove, though his mother had a lifetime interest in the estate.

Ann and Eliza Croghan attended the Domestic Academy near Springfield, Kentucky in 1809. The Croghan twins, Charles and Nicholas, were born June 19, 1802. They attended St. Thomas College, Springfield, Kentucky and Buck Pond Academy near Versailles, Kentucky. Edmund was born September 12, 1805 and attended Jefferson Seminary in Louisville, Kentucky. The last three Croghan sons died young.[26]

CHAPTER SIX

Explorers Leave the Falls

President Thomas Jefferson had a vision twenty years before the Lewis and Clark Expedition began in 1803. He had asked General George Rogers Clark how he would like to lead an expedition to the Pacific. In a letter, dated December 1783, Jefferson expressed deep concern about the large sum of money set aside by England for exploring the country from the Mississippi to California. England pretended it was to promote knowledge, but Jefferson feared the English were planning on colonizing that area. He doubted the United States had enough spirit to raise the money, but if it could, he asked the General, "How would you like to lead such a party?"[1]

The statement "doubted the country America had enough spirit to raise the money" more than likely holds the key to Clark's refusal to accept the offer. He lived daily with the memory of the loss he had suffered in using his fortune for his battles, all for his ungrateful country which refused to reimburse him, leaving him penniless. Another deciding factor was his health, for he lived with bouts of painful rheumatism resulting from exposure to severe weather while serving his country.

He did offer some suggestions if ever a trip were to be made. To Jefferson, he wrote that large parties would never answer the purpose, and would only alarm the Indian nations they passed through. Also, the languages of the distant nations they would pass through would have to be learned.[2]

President Jefferson's relative and secretary, Meriwether Lewis, had served with William Clark under the command of General Anthony Wayne in 1795–1796. The following years 1796–1803, William had made many trips to the East coast. In 1798, he was on the Ohio and Mississippi rivers to New Orleans where he purchased a passport and sailed to New York. He traveled back and forth during the years 1800, 1801 and 1802 between Kentucky and the east. In his words, he had "frequent reasons to visit the

Eastern States & Washington where I became acquainted with President. Mr. Jefferson and was introduced to Jefferson's private secretary."[3]

December 1802, President Jefferson received a letter from General Clark in which he wrote, "I will with greatest pleasure give my bro William every information in my power which may be of Service to your Administration. He is well qualified almost for any business. If it should be in your power to Confer on him any post of Honor and profit in this Country in which we live, it will exceedingly gratify me."[4]

The President heeded the recommendation made by General George Clark. Meriwether Lewis had also expressed his wish to have Clark for his partner on the expedition and the President had expressed "an anxious wish that you would consent to join me in this enterprise."[5]

Lewis informed Clark that he would be recruiting some good hunters, stout, healthy, unmarried men, accustomed to the woods, and capable of bearing bodily fatigue in a pretty considerable degree.[6] Clark found some men who met Lewis' requirements and was waiting for Lewis' approval upon his arrival at the Falls of the Ohio. One important recruit was William's boyhood companion York. This faithful servant, a slave, was mentioned many times in the journals kept by Lewis and Clark.

In 1803, the two were on their way to fame. In preparation for the expedition, Lewis had consulted and listened to those in the field of medicine, collected measuring instruments and other items necessary for collecting specimens and survival.[7] William had more experience on the frontier. From a young age, he had listened to his older brother about survival in the wild untamed country, had learned about Indian traditions, and was both daring and resourceful. William was the map maker and returned with illustrations of birds, fish, and animals in detail.[8]

The need to learn the Indian language was met by engaging the now famous Indian girl Sacagawea as a translator and much has been written about her; but how different the journey would have been without this girl who made the journey there and back with a baby strapped to her back. She guided the explorers through the Rocky Mountains and due to an unexpected event, recognized her brother after many years of separation. He was among a group of warriors, but his help and influence among the Indian tribes along the way resulted in the Indians supplying horses, food and guidance for the explorers.

Meriwether Lewis had scientific skills, but William Clark was not ignorant of nursing skills. He stayed by Sacagawea's side for three days and nights when she was feverish and near death. He made poultices, mixed herbs, and gave her special care until her recovery.[9]

Once during a terrifying storm on the river, the journals in which Lewis and Clark had so carefully kept notes each day went overboard into the water. Sacagawea's quick action of diving into the water and retrieving the expedition journals was a historical act which should be recognized as one of the more important events on the expedition.

In 1900, Eva Emery Dye, historian and writer, was the one who brought attention to Sacagawea's contribution to the success of that monumental event. After two hundred years had passed, Congress finally gave proper recognition to Sacagawea by dedicating a statue of Sacagawea which is placed in the capital rotunda in Washington, D.C.

Many authors have written about the remarkable journey this group of men made through dangerous unknown territory. Miraculously, all but one returned to their loved ones. Lewis and Clark were successful because both were determined to complete this long difficult journey, but most of all they were compatible and had a special respect for each other.

President Jefferson's expectations to find a continuous direct waterway to the Pacific were not met, but it was discovered that traveling the Columbia River was much shorter than Britain's route. They accomplished all the President Jefferson had asked of them.

When the explorers left Missouri it would be two and one-half years before anyone would hear from them and one of the happiest moments in Lucy's life was the safe return of her youngest brother from the expedition. The trip took so long and not knowing what fate might have come their way, was a heavy burden for the family. In the fall of 1806, they returned making their first stop at St. Louis, where the citizens of St. Louis had turned out with a boisterous greeting.

After days of celebrating in St. Louis, they visited with William Harrison at Vincennes, Indiana. Leaving there, they came up the Ohio River to Louisville, and on November 5, 1806, they disembarked at the first sight of home, the lonely cabin above the Ohio River Falls on the Indiana side. There, the returning heroes met General George Rogers Clark, at the Falls of Ohio, the place where the expedition began.

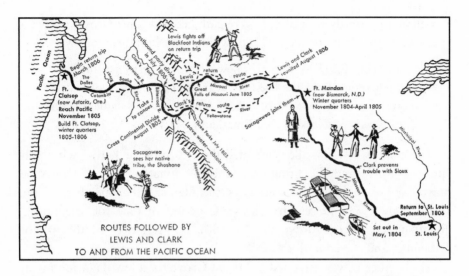

Lewis fights off
Blackfoot Indians
on return trip

Eastbound party divides
in July 1806

Clark's Fork

Lewis

return

route

Lewis and Clark
reunited August 1806

Ft. Mandan
(now Bismarck, N.D.)
Winter quarters
November 1804-April 1805

Clark's
return route

Yellowstone
River

Sacagawea joins them

Missouri River

Great
Falls of Missouri June 1805

Mississippi River

Begin return trip
March 1806

Pacific Ocean

The
Dalles

Columbia

Snake

Clearwater R.

Ft.
Clatsop
(now Astoria, Ore.)
Reach Pacific
November 1805
Build Ft. Clatsop,
winter quarters
1805-1806

Take
to canoes

Cross Continental Divide
August 1805

Three Forks July 1805

Leave water—obtain horses

Rocky Mountains

Sacagawea
sees her native
tribe, the Shoshone

Clark prevents
trouble with Sioux

Missouri
River

Return to St. Louis
September 1806

Set out in
May, 1804

St. Louis

ROUTES FOLLOWED BY
LEWIS AND CLARK
TO AND FROM THE PACIFIC OCEAN

Previous to the return of the explorers, General Clark had been follow-ing the pursuits of Napoleon in Europe, but great things of a different nature were happening here in this new nation such as returning from the great Pacific.

William Clark kept a promise made to his brother-in-law Major William Croghan two years before he left on the long journey:

> William Clark, St. Charles May 21, 1804 to William Croghan [Louisville].
>
> My friend Capt. Lewis expressed some sorrow that you happened not to be home at the time he passed down, but hopes to see you on his return to the U. States, as to my sel, I have, do, and shall [al] ways have that Brotherly affection for you which you are well assured I always possessed, and hope that in less than two years to see you & that family (yours) whome I have every effection for, at your own house.[10]

In much need of a new wardrobe, the heroes crossed over to the Louisville side where merchants Fitzhugh and Gwathmey were waiting to dress the returning heroes in the finest attire for gentlemen, complete with ruffles and buckled shoes.[11] They were home for celebrating. York found this amusing when Clark had shed his buckskins for ruffles and buckled shoes.[12]

General Clark accompanied the heroes to Locust Grove, which was ablaze that night with welcome and the Clark brothers, sisters and Major

Croghan, the courtly host of old, were gathered at Locust Grove.[13] What a greeting was waiting for them! There would have been tears of joy.

> November 5th, 1806, Captains Lewis & Clark arrived at the Falls
> on their return from the Pacific Ocean after an absence of a little
> more than three years. 6 & 7 & 8 rain at Majr. Croghans with
> Capt. Lewis & Clark. (From Jonathan's Diary)

The citizens of Louisville celebrated the return of the heroes with banquets, bonfires and the firing of cannons.[14] Jonathan's diary shows he stayed at Locust Grove for four days and nights. Other members of the Clark family were present for the same period of time for celebrating. Jonathan left Locust Grove on the 10th, but returned on the 4th of December and stayed until the 15th.

Meriwether Lewis remained at Locust Grove for a few days, joining in the celebrating. Of all the famous and distinguished guests who sat at the Locust Grove dining room table, none would have furnished more stimulating conversation than Lewis and Clark with their exciting stories of adventure.

When Lewis and Clark arrived at Locust Grove, Clark would have proudly displayed his artifacts, including skins and furs, which dried in Chouteau's big fur warehouse while in St. Louis. Clark's brothers and sisters were the most important people in his life, and sharing these trophies with those he loved was a big moment in his life. He would have enjoyed every minute while they looked on in amazement and disbelief. William's prized relics would not have remained packed in boxes when he returned to Locust Grove and the ballroom floor was the best place to spread the prized artifacts.

> Spread around for exhibition were Mandan robes, fleeces of the
> mountain goat, Clatsop hats, buffalo horns, and Indian baskets,
> Captain Clark's 'tiger-cut coat", Indian curios, and skins of griz-
> zly bears,—each article suggestive of adventure.[15]

Jonathan Clark wrote to brother Edmund with all the news about the night of William's return to Locust Grove, which included the line, "artifacts were spread on the ballroom floor." Eva Dye had copied the letter written by Jonathan to Edmund Clark from Isaac Clark's collection. The original was left at Mulberry Hill where Edmund Clark died in 1815. There were no restrictions on the Dye collection, therefore, all of Dye's original collection are missing from the large envelope that held them at Oregon Historical Society history center. "Found in the Dye file, there is one letter from William Hancock Clark (1901) requesting a copy of that particular story from Dye. So they had discussed it, but perhaps in person when Dye visited Clark in New York."[16]

The following is a list of articles forwarded to Louisville by Captain Clark in care of Mr. Wolpards:

William Clark
From St. Louis 1806

One large Box Containing
    4 large Horns of the Bighorn animal
    2 Sceletens do   do do
    2 Skins horns & bons of do
    4 Mandan Robes of Buffalow
    1 Indian Blanket of the Sheep
    1 Sheep Skin of the rocky mountains
    1 Brarow Skin
    3 Bear Skins of the White Speces
    barking squirls
    2 Skins of the big horn
    1 Hat made by the Clatsops Indians
    2 Indian Baskets
    4 buffalow horns
    1 long box of sundery articles
    2 Tin box containing Medicine &c &c. &cc
    a Small Box of papers
    Books and Sundery Small articles
    a Hat Box containing the 4 vols of the Deckinsery of arts
        and ciences
    two indian wallets a tale of the black taile Deer of the Ocean
        & a Vulters quill with a buffalow Coat.

The articles were sent from St. Louis to Louisville by Adam Woolford (or Woodford). He was the one who operated a boat between St. Louis and Louisville.[17]

49

In William Clark's will, he left his expedition relics to his first son, Meriwether Lewis Clark. The following information is found in family letters written by William Clark's grandson William Hancock Clark to Eva Emery Dye in 1900:

William Hancock Clark's father [Meriwether Lewis Clark] had left the Lewis and Clark relics to youngest brother, Jefferson K. Clark, who had adopted his [MLC's] youngest son Jefferson Clark. "They were left in trust for Meriwether's children, they thought." But in 1893, a codicil was added to his will leaving everything real and personal to his [Jefferson K. Clark] widow, Lena Jacobs Clark.[18] In 1905, these priceless relics had not been returned to their rightful owners. There were some legal steps taken, but without success.

Meriwether Lewis's relics were with a delegation of Osage Indians who had left St. Louis, with a pack train loaded with the party's baggage and whatever plants, seeds, dried skins, animal skeletons, and furs had not been ruined in water.

After a few days of celebrating in Louisville, Meriwether departed for Virginia, leaving Clark with his family. Meriwether's first desire was to see his mother. Before he left on the expedition, his mother was living in Georgia with her second husband, Colonel Marks. Regretting his inability to see her before he embarked on the trip, he tried to alleviate her uneasiness about himself by writing the following:

> The nature of this Expedition is by no means dangerous, my route will be altogether through tribes of Indians who are perfectly friendly to the United States. The charge of this Expedition is honorable to myself, as it is important to my country. I truly hope that you will not suffer yourself to indulge any anxiety for my Safety and promising to write.[19]

Two and one-half years had passed since he had heard from his mother and had no way of knowing whether his mother was still living or no longer there. In St. Louis, Meriwether had sent a hastily scrawled note to President Jefferson, "I am very anxious to learn the state of my friends in Albemarle particularly whether my mother is yet living."[20]

Lewis was in Frankfort, Kentucky, in November and on the 15 with Big White (Indian escort) took the Old Wilderness Road for Washington, only stopping in Virginia to be with the Lewis family for a happy homecoming. The pack train had left Louisville at the same time Meriwether Lewis left.

Chouteau and the Osage chiefs proceeded on with pack horses via Lexington, arriving in Washington on Christmas Eve. Jefferson wrote Charles Willson Peale informing him that he expected Lewis daily and would be bringing specimens to the American Philosophical Society.[21]

The Croghan boys, O'Fallons and the Andersons were away at school and missed all the excitement. William Clark wrote to Major Croghan on December 14, expressing his regrets that it wasn't in his power to visit Locust Grove before he left. He intended to proceed as far as Colonel Anderson's, then to Danville on his way, where he would see the boys.[22] November 20th, 1806, Jonathan Clark received a letter from his daughter, Eleanor Ettings Temple and husband Benjamin from Lexington, in which they expressed a wish that William Clark would call on them before going east.

Lewis had arrived and reported to President Thomas Jefferson, one week before Clark arrived. The returning heroes experienced exhilarating fame and celebrations wherever they were. Dancing with these two young famous and successful celebrated heroes would have been an event remembered by the ladies in each town for years to come. William wrote to brother Edmund that he had become quite gallant dancing with the ladies, but William was in love with a daughter of Colonel George Hancock, a member of Washington's Fourth Congress.[23]

Clark shortened his stay with the President because he was anxious to return to Julia in Virginia and ask for her hand in marriage. His proposal of marriage was readily accepted by Julia and thirty seven year old Clark returned to Louisville, to prepare for his upcoming marriage to the sixteen year old Julia, a first cousin of Zachary Taylor. She was also related to President Madison and his wife, Dolly.

It has been speculated that it is doubtful William had romantic intentions for Julia before the Northwest Expedition as she was only twelve. Well, Clark must have fallen in love with that twelve-year old girl, after all, he named the Judith River for her, which has been accepted as fact.

To Eva Emery Dye. Mrs. Mary Kennerly Taylor wrote the following:

> Uncle-Clark has often told me of his first meeting with Julia and Harriett, they were nearly the same age. Julia a few months older than Harriett. [younger] The girls were both riding on one horse & trying their best to get the horse to go, both using switches to urge him on, but he would not move. Clark riding by, the girls

Governor William Clark, ca. 1810, painting by Charles Willson Peale (courtesy of Independence National Historical Park)

Julia Hancock Clark, William Clark's first wife. (Courtesy of the Missouri Historical Society)

Harriet Kennerly Radford Clark, William Clark's second wife. From W.C. Kennerly, Persimmon Hill.

called on him to help them. He helped the children, leading the horse for some time & when he mounted his own theirs followed. This was his first introduction to the girls. I have often heard him speak of it & say though a very young man he made up his mind that Julia should be his wife.[24]

The girls, about the same age, were both beautiful and accomplished musicians on harp, guitar and piano. Harriet would later become Clark's second wife.

William Clark and Julia Hancock were married at the Hancock mansion, *Fotheringay*, January 5, 1808, in the beautiful drawing room. President Jefferson sent a wedding gift to the new bride to be presented by Clark. When she opened the box, inside was a set of topaz and pearl jewelry, said to consist of seven pieces. Later, this jewelry would be worn by the wife of their eldest son, Meriwether Lewis Clark, and the wife of his son, Meriwether Lewis Clark, Jr.

The famous wedding jewelry is now in the collection of the Missouri Historical Society. Jefferson's treasured collections had a way of going on long journeys of twists and turns before finding their final resting places. Unlike the Lewis artifacts, which were brought back for the President and ended up in Chesapeake Bay, the jewelry survived wide travels. The expensive jewelry with historic value was given to a young lady in Europe sometime around 1900.

The following letter was written to Orville Anderson Tyler at Short Hills, New Jersey, by H. Armour Smith, Yonkers, New York, dated September 14, 1935.

> It occurred to me that you might be interested in a set of topaz and pearl jewelry that was presented to Julia Hancock Clark upon her marriage to General William Clark at Fincastle, Virginia, January 5, 1808, by President Thomas Jefferson. The jewelry was also worn by the wife of their eldest son, Meriwether Lewis Clark and the wife of their eldest son, William Hancock Clark. The jewelry is now in Europe, having been presented to a Miss Egholm by Major William Hancock Clark. Miss Egholm has since died and the jewelry has been offered to me by her heir.
>
> This set of historic jewelry has been in Europe for a number of years and the present owner, being aware of its value would like to sell it.

I have purchased from him some rather interesting things former-
ly the property of Major Clark.(?) I have recently received a photograph of the jewelry which consists of seven pieces, apparently
still in the original case.

Should you be interested—I will be glad to have you see the photograph and give you what further information I may have.[25]

After much correspondence and exchange of money, the set was
returned to St. Louis, Missouri, where the bride and groom had first
arrived with the jewels in 1808.

William Clark and his bride spent the late winter in the eastern cities
and the opening of spring. They began their wedding trip down the Ohio
River. Five maids traveled with them to take care of Julia and her personal
belongings.[26] Of course there was a stop at Locust Grove for much celebrating.

In those days, the Clark family took every opportunity to celebrate,
and they didn't let this one pass. In May 1808, Edmund was living down
the Ohio River in Henderson County, and received a letter from Jonathan
describing the great celebration that took place in honor of the new bride
and groom.

In May, Edmund had just learned about his brother William's marriage, but he could not find the first name of the new bride. "Drink good
eating was prepared to welcome them, I am sure and twas a pity that I
miss'd all the good dances. I make no doubt there was much joy and, who
knows had some sorrow too, for that is frequent among a multitude."[27]

Jonathan's Diary, "... went to Shipping port to see the wedding party
off".

Traveling down the river, a number of gentlemen and ladies joined the
newlyweds. Major Croghan, George Croghan and W. Richard Brown were
among the many going as far as the mouth of the Cumberland. George
Rogers Clark Sullivan said he "... supposes, there will be great crying, when
they part."[28] Upon reaching the mouth of the Cumberland, William
Croghan probably took the wedding party on a tour of Smithland, the
town he had established.

Meriwether Lewis described the house he found for the newlyweds in St.
Louis as having four rooms on the ground floor, an attic for slaves, and a
garden already planted. Lewis had been appointed Missouri Governor and

went to great pains to ensure his friend's bride a pleasant voyage to St. Louis. Clark was bringing furniture, heavy equipment for a horse-power mill, blacksmith tools, and other bulky articles in two large keelboats.[29]

For his old friend and bride, Governor Lewis sent the services of a Regular army ensign and a military guard to meet the party at the mouth of the river. Lewis wanted the ladies to be comfortable on a keelboat, "I trust, the Governor added, you do not mean merely to tantalize us by the promise you have made of bringing with you some of your Nieces, I have already flattered the community of St. Louis with this valuable acquisition to our female society."[30]

Ann Anderson was the only niece who went to St. Louis with her Uncle William and his new bride on that boat. The beautiful and accomplished Miss Anderson was the daughter of William's sister, Elizabeth Clark Anderson, who died when Ann was an infant. Clark's niece was said to be a pretty girl and caused a flutter in society of old St. Louis.[31]

Before William Clark moved to Missouri, he had one more bit of important business to take care of. Upon receiving an invitation to attend the Centennial Celebration for the Lewis and Clark Expedition in St. Louis, the following letter was written and is self-explanatory.

3103 Law Ave., St. Louis, Dec. 26, 1900

Mrs. Eva Dye Oregon City

Dear Madam

The William Clark of the Lewis and Clark Expedition sustained to my father, the double relation of Master and Father.

When he moved to Missouri, he emancipated his mulatto children, sent them to Ohio and made some provision for their care and education.

I was born in Ohio, not Missouri.

We, I speak for myself and brothers and sister, have never cared to claim Kinship with the white branch of the family, nor do I now, court any notoriety in that connection.

Yrs.
Pete H. Clark

P.S. One of those Clarks died in New York a few months ago. He was brought here for burial. The service even held in the Cathedral (Episcopal) on 13 near Olive St. A note addressed to the dean of that church will probably elicit the information you desire. P.H.C.[32]

Surnames for slaves were rarely used, since a surname would have been too easy to trace, revealing who the white father was, usually the master of the estate. Also, a surname would only be a reminder that they were once a family torn apart and sold at the auction block. Ohio was a free state. In Kentucky, emancipation of slaves required a court order and William would have been forced to admit fathering these children.

The slave York and William Clark had been inseparable since they were small boys, and this close relationship continued on the western expedition. Refusing to free York, for some unknown reason in 1810, William sent York back to Kentucky where he was hired out and fell on hard times. William hired a man named Rooskoski from Poland, who filled the position of valet to Clark.[33] Nephew John O'Fallon appealed to William Clark to rescue York from the mistreatment of a Mr. Young, but nothing was done for York.[34]

Did Julia Hancock have anything to do with William sending York away? There is a possibility she may have influenced her husband. In Virginia the Hancocks were proud of their well trained slaves who were considered to be superior to others.[35] Julia may also have resented her husband's close relationship with York.

While William Clark was in Washington celebrating, President Jefferson had asked him to collect more bones at Big Bone Lick in Boone County, Kentucky, and in Louisiana. After returning to the Falls of Ohio, William employed ten helpers for several weeks and collected about 300 bones which were left with his brother George at his cabin. In December 1807, the President wrote to General Clark, instructing him, "*I think your bro might have left by now ... have the bones packed and forwarded for me to William Brown Collection at New Orleans who will send them on to me.*"[36] Upon discovering the Kentucky bones were duplicates of ones brought from the West, they were sent to a museum in France.

The expedition artifacts and specimens sent to President Jefferson by Meriwether Lewis were delivered by mule packs to Washington, but when President Jefferson's presidential term ended, he was anxious to return to Monticello. He discovered it was no small job to pack and send the large artifacts. Charles Wilson Peale mounted the head and horns of the big horn sheep to be shipped down the Chesapeake for the hall at Monticello. Jefferson in turn sent heavy things in the same boat (including groceries and household goods) to Richmond by water. The boat sank and all artifacts from the expedition were lost, except the horns.

The incident had to be reported to Meriwether Lewis but, informing Lewis was a painful task for the president since he had received twenty-five boxes of artifacts brought back from the West and almost everything was lost. Lewis let it be known that he sincerely regretted the loss, and replied to Jefferson with a touch of criticism at the lack of care of the priceless treasures. "It seems peculiarly unfortunate that those at least, which had passed the continent of America and after their exposure to so many casualties and wrisks should have met such destiny in their passage through as small portion only of the Chesapeak."[37]

A famous legend has it that a Monticello servant burned the Lewis and Clark artifacts and specimens as trash. Checking the source and validity of this story, Monticello historians have discounted its validity, but even today after two hundred years, visitors to Monticello continue to ask for the story to be told. The false story of burning artifacts could have been planted to cover up the information that the final resting place of the specimens is the bottom of the Chesapeake River.

(This letter was found among Clark's papers owned by his granddaughters in New York City, and was published in the century. A handsome tablet of solid brass 8 x 12 inches in size, engraved as follows:

SACAJAWEA
Guide to Lewis and Clark Expedition
1804–1806
Identified by Rev. John Roberts, who officiated at her burial,
April 21, 1884.

This tablet to Sacajawea, the Boat Launcher, was placed over the grave of Sacajawea at the Wind River U.S. Indian School, at Wind River, Wyoming in 1909.

CHAPTER SEVEN

General George Rogers Clark

While others seemed to be absorbed in different pursuits, George Rogers Clark was living alone at his cabin. He asked his nephew George Gwathmey, a merchant across the river in Louisville, to send him some books. Gwathmey sent *The Life of Frederick, The II*, but, "supposes he has read this." General Clark's nephew would try, that afternoon, to get more books for him.[1] George Rogers once had a very fine library, but books loaned from his collection were never returned.

In 1808, there was much interest in Napoleon Bonaparte. Some feared Napoleon would come over to America and conquer. Thomas Jefferson wasn't swayed by this fear. He was of the opinion that before taking Spain and Portugal, Napoleon had to subdue England and Russia. General George was anxious for news of Napoleon's fate.

After expedition celebrations had ended and William was settled in St. Louis with his young family, the Clark brothers George and William would never have the same close bond that once existed. They had always shared common interests such as military affairs and a fascination for natural history, especially in the collection and study of extinct animals. In front of George's cabin, there were the petrified specimens of the vertebrae and tusks of the "mammoth," as he called it. Petrified fish and terrapins were part of the collection. William was familiar with the petrified backbone of a mammoth which sat outside the front door and was sometimes used by George Rogers as a seat.[2]

Summer of 1808, William and Julia, along with their wedding party, had passed through Louisville. While there, William visited his brother George at the Point of Rock across the river. Indications are that the general's reception of his brother was cool, which didn't surprise their brother Edmund. In a letter to Jonathan from Henderson County [Kentucky] he expresses his thoughts concerning the strained relationship between brothers George and

*General George Rogers Clark, by Matthew Harris Jouett
(courtesy of The Filson Historical Society, Louisville, KY)*

William. "For myself I am glad as I think he (Brother) has stood as a Target long enough and now as a matter in course he must mind his own Business which I hope will turn out to his satisfaction. I held up Brother [G.R.] as long as I could. As to the reception, Bro William received from him was much in the stile I expected. Yr. Afftc Brother, Edmund Clark."[3]

The General was a bitter man at that time, and some of the blame for this rested on William. Five years before this visit, George's last request to the United States government in 1803 was turned down, even though he had made it possible for America to claim half of the nation. All the wealth and good living his brothers were enjoying had been denied him.

> Like the blind Belisarius, he was a dependant
> for his daily bread in the empire he had saved.[4]

In 1803, immediately after Meriwether Lewis and William Clark departed for their expedition, General George Clark wrote to Senator Breckenridge petitioning the government one last time. His petition asked for land equal in quantity to the tract the Piankeshaws had conveyed to him in 1770, but to which he had always disclaimed title. The petition, after setting forth the history of the Indian grant, read:

> My reason for not soliciting Congress before this, was the great Number of petitions before them, and the prospect I yet had of a future support; [meaning by Virginia's payment of his claims] but those prospects are vanished. I engaged in the Revolution with all the Ardour that Youth could possess. My Zeal and Ambition rose with my success, determined to save those frontiers which had been the seat of my toil, at the hazard of my life and fortune.

> At the most gloomy period of the War, when a Ration could not be purchased on Public Credit, I risked my own, gave my Bonds, Mortgaged my lands for supplies, paid strict attention to every department, flattered the friendly and confused the hostile tribes of Indians by my emissaries baffled my internal enemies (the most dangerous of the whole to public interest and carried my point. Thus at the end of the War I had the pleasure of seeing my Country secure, but with the loss of my Manual activity, and a prospect of future indigence demands of very great amount were not paid, others with depreciated Paper Suits commenced against me for those sums in Specie. My Military and other lands, earned by my service, as far as

they would extend were appropriated for the payment of those debts, and demands yet remaining to a considerable amount more than the remains of a shattered fortune will pay. This is truly my situation I see no other resource remaining but to make application to my Country for redress hoping that they will so far ratify the Grant as to allow to your Memorialist an equal quantity of land now the property of the United States or such other relief as may seem proper.[5]

It wasn't a gift he was asking for, but payment of the debt owed him. His petition was rejected. Not one acre was he allowed in the vast territory in which his brothers were reaping great rewards. He had no regrets for the sacrifices he was called upon to make during the Revolution. If he could only get away, he would not have to face reminders of how well everyone around him was living in comfort while he was left with bare necessities.[6]

Years before, General Clark had traveled to his Virginia home in 1783 to be with his family and near his dying brother John. However, another reason for his return to Virginia was to visit Richmond with his original account books and vouchers for his expenses incurred during his western campaign. These accounts had been carefully kept and in order. He was owed four years back pay, but received little cash, since both Virginia and the federal government claimed the other was responsible for payment. Therefore, General Clark was held personally responsible for debts owed to creditors incurred against the state during the Revolutionary War.

When John Clark's family arrived at Mulberry Hill in 1785, General George Clark's father had to struggle to keep him out of debtor's prison. Everything the General owned was lost to his creditors. He could not accept an inheritance from his father's will because the creditors would have taken it. This was a time in history that the state of Virginia and the nation should not be proud for the unspeakable treatment of an old hero that gave the nation half a continent.

Again in 1787, General Clark was asked to present his account books and vouchers since officials of the government could not find the ones he left with them in 1783, they declared them lost. The powers in Virginia had no time to search for them. One hundred and thirty-one years after he received the auditor's receipt for his vouchers—they were found in a room of the auditor's building in Richmond, seventy large packages of those original vouchers. They had not been destroyed by Arnold's British Soldiers.[7]

William Clark wrote to Edmund late in 1797, "I have rode for Brother George in this past year 3,000 miles continually on the pad attempting to save him and have been serviceable in several instances. Got Spanish suit dismissed, and secured his property on that side of the Ohio."[8] William inherited Mulberry Hill and would convince others that he sold it to liquidate some of his brother's debts, but didn't mention the fact that he received a good return for his time spent in collecting those funds and would remind others that he deserved a high commission for acting as a financial agent.[9]

William Clark's 3,000 miles of 'years on the pad', weren't exactly for General Clark's benefit. On one trip in 1798, he traveled from Kentucky on boat to New Orleans and then to the Eastern States by boat around the Southern Coast, and visiting several places on the East Coast. Washington was a place where he frequently visited during the next two years and became "acquainted with the Presidt. Mr Jefferson." He was introduced by his friend Meriwether Lewis, who also made him acquainted with President Jefferson's private secretary.[10]

Before William was asked to go west on the expedition, he and George were in the process of developing water power near George's cabin in connection with the town of Clarksville on the Indiana side. George applied to the trustees of the town for the inherent rights of the front lots on the Ohio and asked that these rights be exclusively granted to his brother William. A canal would be constructed, allowing the passage of boats, as well as serving as a source of supply to water-wheels adjacent, which would provide power for mills and factories. The estimated cost was $150,000.[11]

William was not a man of means which suggests they expected the financing would come from the sale of Mulberry Hill and gaining possession of George's land in Kentucky and Indiana. On the Ohio River, he gave William 74,000 acres in cancellation of a $2,100 debt, 15,000 acres in payment for a debt and 2,000 acres for William's services during this time.[12]

William Clark also inherited from his father twenty-four slaves, a distillery, a grist mill and 7,040 acres of Kentucky land. George had signed over to his father, lands north of the Ohio River, which William inherited. General George Rogers Clark received only three slaves from his father's entire estate. Plans for constructing a canal and supplying water power on the Indiana side of the river were interrupted when President Jefferson appointed William Clark to join Meriwether for the expedition.

Jonathan Clark moved to Kentucky in 1802 and lived at Trough Springs near Louisville. His father-in-law, Isaac Hite, Sr. was one of the wealthiest men in Virginia and was recognized as a wealthy land owner in Mercer County, Kentucky. Jonathan devoted himself to business with great success, accumulating a large fortune in real estate, as well as personal property. The inventory of the latter, returned by administrators, Abraham Hite, his wife's cousin; and John H. Clark, his son, covers eleven pages in the book of inventories in Jefferson County, Kentucky.[13]

Brother Edmund Clark sold his business as a merchant in Richmond, Virginia, and settled on his place in Henderson, Kentucky. He also was into land speculation in Western Kentucky, searching for more property for himself and for Jonathan to purchase.[14]

After William Clark's return from the western journey he had been appointed Commissioner of Indian Affairs, a commission once desired by George for himself. William later became Governor of Missouri, living in fine housing, formal gardens, with all the comforts of life as well as honor and respect. When the party left Louisville, "William was traveling with two keelboats, bringing his furniture and heavy equipment for the Indians, including a horse-power mill and blacksmith's tools."[15] This equipment could have been the same ones which were to be used for the water power project in 1802. It is easy to assume, this was cause for the cool reception William received from his brother.

Life was good for three of the brothers, while at the same time, General George Clark was fighting failing health, bitterness and alcohol. He lived in poverty and obscurity. Rheumatism had gotten the best of him which caused limping so badly, a cane was necessary for walking. His poor health was due to exposure to the extreme elements as a warrior. The following is a detailed description of the clothing worn by Clark and his men during the war:

> The hunting shirt was generally made of linsey, sometimes of course linen, and a few of dressed deer skins. These last were very cold and uncomfortable in wet weather. The shirt and jacket were of the common fashion. A pair of drawers or breeches and leggins, were the dress of the thighs and legs; a pair of moccasins answered for the feet much better than shoes.

> Moccasins in ordinary use cost but a few hours labor to make them. This was done by an instrument denominated a moccasin awl, which was made of the back spring of an old clasp knife. This awl

with its buck-horn handle, was an appendage of every shot pouch strap, together with a roll of buckskin for mending the moccasins. This was the labor of almost every evening. They were sewed together and patched with deerskin thongs, or whangs as they were commonly called.

In cold weather the moccasins were well stuffed with deer's hair, or dry leaves so as to keep the feet comfortably warm; but in wet weather it was usually said that wearing them was 'a decent way of going barefooted; and such was the fact, owing to the spongy texture of the leather of which they were made.

Owing to this defective covering of the feet, more than to any other circumstance, the greater number of our warriors and hunters were afflicted with the rheumatism in their limbs. Of this disease they were all apprehensive in cold or wet weather, and therefore always slept with their feet to the fire to prevent or cure it as well as they could. This practice unquestionably had a very salutary effect, and prevented many of them from becoming confirmed cripples in early life.[16]

Before 1779, the Kentucky frontier was in desperate need for protection and only Clark was the one who spoke up for their protection during the Revolutionary War. British Officer Henry Hamilton was sending Indians down to Kentucky, burning the settlers' homes and returning with their scalps to be paid by Hamilton. This British General was appropriately called "The Hairbuyer." The military and government leaders were too close to the eastern battlefields to hear the pleas of a defenseless few out in Kentucky. So, Clark organized, planned, led, and even financed his campaign almost on his own using his personal fortune to underwrite his conquest of the Northwest.

Almost alone, he won the Northwest Empire. The march on Vincennes is one of the greatest military acts of the Revolutionary War. Secrecy and surprise were the elements maneuvered for success. They hid their canoes so no tracks would be left. With 175 frontiersmen, he trudged almost 200 miles through desperate cold, breaking ice to wade the rivers, with water up to their waist to seize the British stronghold of the west, Post Vincennes on the Wabash River. They had marched from Corn Island at the Falls of the Ohio (now Louisville) in the cold month of February 1779. It was incredible how Clark and his soldiers surprised Hamilton, captured him and sent him back East as a prisoner.

In this one, bold stroke, George Rogers Clark broke the power of the British Northwest and added half a continent to our young republic.—Janet Walker.

George Rogers Clark's relentless determination to march on Vincennes in the month of February for ending Hamilton's treacherous deeds delivered to the settlers in Kentucky was never properly repaid or rewarded. George was the only one who had the steel and commitment to end this atrocity that was occurring on the frontier.

The following list was found in baggage of the English army which had been sent to British Governor of Canada, Colonel Hamilton.

England's Humanity: It is generally known that savages were employed by the King of England, George III, and paid at so much per scalp of man, woman, and child during the Revolutionary War. A few items from this terrible trade in human flesh may interest some of our readers, and show how this paternal King strove to crush out the noble spirit of independence of our early heroes. Here is a list of a number of packages that were sent by one James Boyd, from a Captain Crawford, to the British Governor of Canada, Colonel Hamilton. These packages of scalps were found among the baggage of the English army, after the defeat of Burgoyne, cured and dried, with Indian marks upon them.

Package 1
Containing forty-three scalps of Congress soldiers, killed in different skirmishes, stretched on black hoops four inches in diameter.

Package 2
Containing ninety-eight farmers' scalps, killed in their houses, on red hoops, with a figure of a hoe painted on each to denote their occupation.

Package 3
Containing one hundred and two farmers' scalps, eighteen of them marked with yellow flames, to signify that they were burned alive.

Package 4
Containing eighty-eight scalps of women, hair long and braided, to show that they were mothers.

Package 5
Containing one hundred and ninety-three scalps of boys of various ages, on small green hoops.

Package 6
Two hundred and eleven girls's scalps, big and little, on small yellow hoops.

Such was the stuff on which English royalists were made. Boston Post THFC[17]

In 1783, the British surrendered to the colonists. The Treaty of Paris was negotiated by Franklin, Adams, and John Jay who capitalized on the conquests of General George Rogers Clark in the "Old Northwest", adding Ohio, Illinois, Indiana, Wisconsin and Michigan to the young nation.

General G.R. Clark's sixty-six year life span cannot be compared with the longevity of Daniel Boone or other warriors. In 1809, Clark was very feeble and crippled and as a result, fell into the fireplace. His leg was burned so badly, it had to be amputated. Family letters refer to his uncontrollable tremors before falling into the fire and the amputation which took place.

> He burned a small place on the side of his knee. It was always supposed that he was intoxicated, but it was not [certainly] known. It was at first considered by him as a small matter, but until his cook observed the inflamed state seen that more skillful attention was called to it. He was at the time this occurred in a feeble state of health and tottered when he walked and his fall may have been for this reason, but there is not doubt in her recollection—at this time his habits were intemperate.

> From neglect, the place inflamed & his leg became swollen until it (illegible) the attention of a negro woman in his service who communicated her conviction that the matter needed the attention of a physician & when it was examined, it was found that mortification had (illegible) that amputation was indefensible. When Amputation was performed he had become so weakened by disease and the affects of his (illegible), that it was feared he would not survive the operation.[18]

The following was written by Ann Croghan Jesup's granddaughter, Violet Blair Janin:

Major Croghan stood beside him as the Doctor was operating and put his hand on him for fear he might move. GRC smiled up at him and said, Don't be afraid, Major, I will keep still. Mother [Mary Jesup Blair —1825–1914] said her Mother [Ann Croghan Jesup] was a child in the house at the time. He died when my grandmother was about 20 in 1818. She was Ann Heron Croghan, you know, and she married in May 1822 Gen. Thomas Sidney Jesup.[19]

Ann Croghan, eleven years old at the time, could have been with her parents at the time of the amputation, but it would have been in her Aunt Frances Fitzhugh's home in downtown Louisville.

It has been repeated that there were no anesthetics, but contrary to reports that Clark had nothing to ease the pain, some letters and sources reveal otherwise.[20] From all indications, the report of no anesthetics was repeated at the family's request. Jonathan Clark's wife, Sarah Hite Clark was the only one present in the room during the amputation. Her memory of events that took place during the amputation differs, somewhat, from this popular version.

The room used for amputation was full, but as the blood began to flow, the room began to become empty of its occupants. Dr. Fergurson asked that someone should stay to assist him. General G.R. Clark spoke up and asked for his sister-in-law, Sarah Hite Clark, Jonathan's wife, to stay. She bathed Clark's brow and later would tell how she almost didn't make it from becoming nauseous as the others had before and it was doubtful if she would be able to stay. Dr. Fergurson and Sarah Hite would respond to his grimacing and grinding of teeth when the pain became almost unbearable. At one time, Clark said, "Dr. you hurt me," by which the doctor said, "the knife got caught in some skin, but it would not happen, again."[21] "Music was played by a fife and drum band, while marching around outside." One reason given for this was so his cries of pain would not be heard during the operation. Sarah Hite Clark remembered the circumstance under which the amputation took place.

It is hard to imagine there were no cries of pain when the surgeon's knife cut round to the bone, used a saw to sever the bone and a red hot iron to cauterize the open raw flesh after the amputation.[22] George Rogers was given whiskey, the only anesthetic used to dull pain in those days. Years later, Dr. Ferguson's son wrote that the family did not want the use of whiskey mentioned out of respect for the "old patriot."[23] General Clark always praised his sister-in-law and never forgot to say that she was a good woman.

During the surgery, the General's Sister Fanny walked up and down outside the room grieving and later decided that she wanted her brother to live with her, because she was the youngest. But, with his youngest sister he would have been living on a second floor above her husband's store. Lucy and William Croghan took her brother to their comfortable home in the country. The library on the first floor at Locust Grove was converted into a bedroom for the old General and Kitt was near at all times.

A young niece of General Clark, Ann Thruston Farrar, recalls going with her mother, Fanny, to visit General Clark at Point of Rock, across the Ohio River. On a lonely bluff on the Indiana side, Old Uncle Kitt, his devoted slave, was always in attendance.

Ann Thruston Farrar was in her parents' home during the operation and wrote the following:

> He had a nice little cottage and farmed it on a small scale (for the watermelons were glorious). Later, he was brought to my mother's and stepfather's, Judge Fitzhugh, in Louisville, when his leg was amputated, I think by Dr. Ferguson, and suppose others, for the house seemed full. He never walked afterwards. A band of music playing all the time. The array of instruments attracted my curiosity with my mother walking up and down listening at the door in the deepest grief. Often, I would hear her and Pa speak of his fortitude and patience under the operation.
>
> I know how long he was with us, only that Ma was grieved at this removal to Uncle Croghan's, feeling, I suppose, that she was the youngest of the family and ought to have the care of him. She was a devoted sister, and would beg of Uncle Kitt to take good care of her dear afflicted brother.[24]

Family letters indicate that during the last nine years of General Clark's life, he was showered with loving care while living with his old friend Major Croghan and his sister Lucy. The long porch on the river side of the house was his favorite place to spend long hours surveying the comings and goings of family and friends.

CHAPTER EIGHT

1811 Earthquake and River Trade

The days went rather smoothly for about three years after the return of William Clark from the expedition and General Clark's survival of leg amputation and then, in 1811, an alarming event took place at the banks of the Ohio in Louisville. One still moonlit night in October citizens were alarmed by an extraordinary sound of a steam whistle and engine coming down the river. It was Fulton's steamboat, the *Orleans*, which had come from Pittsburgh. The excitement, mixed with terror and surprise brought out many in town from their beds to see what was happening, some thinking a comet had fallen.[1] The coming of the steamboat marked a great change in travel. Travel by stagecoach with overnight stops at wayside taverns, now became a second choice to travelers.

The excitement which took place on the arrival of the steamboat wasn't the only stirring news to hit Louisville that fall. Another event of a different nature was the shocking news of Jonathan Clark's death which was upsetting for all citizens of Louisville. Jonathan's nephew, John O'Fallon wrote that while he was on the Tippecanoe expedition with General Harrison at Vincennes, he received the bad news that Uncle Jonathan had gotten wet while witnessing the launching of a schooner, took cold and died on November 24, 1811 in an apoplectic seizure.[2] The Clark family had looked to him as a stabilizing force in their lives.

From General George Clark, he remarked: "It was hard that he, who wanted to live, should die, while I, who wanted to die, should live."[3]

With the coming of the steamboat, the year ended with the devil himself. A series of major earthquakes took place in Western Tennessee and Kentucky with tremors lasting through 1812. No mention of the quake is made in Croghan letters, but it was alarming to everyone. The houses in Louisville had two pendulums suspended from their ceilings, such as were found in every house at that time. One was heavier and longer in length, the other was shorter and lighter in weight and would start swinging at the first light rumblings of a quake. If the heavier pendulum started swinging radically, this indicated a major quake that could cause windows to break and the collapse of buildings. They were used so people could run from their houses and stand in the streets to escape falling buildings until things were still once more.[4]

The worst of these quakes caused the Ohio and Mississippi rivers to flow backwards for a while. Reelfoot Lake in Tennessee was formed by the eruptions and rolling of the earth at that time. " . . . upon the River, seven Boats passing on the falls today; some with and some without crews on boards. Much howling and lamentation were heard from a boat entering the falls this night, voices of men women and children."[5] People "got religion" and all churches were full for a while. Then slowly as life went back to normal, card games and alcohol reappeared.

Well-built brick houses withstood the impact of the earthquakes much better than others.

Colonel Anderson, Croghan's brother-in-law, discovered that Soldiers Retreat would not fare so well. The house was built using the same floor plan as Locust Grove, but on a grander scale, with massive double doors at the entrance. The super-abundant gray limestone of the Beargrass region was used instead of bricks for the Anderson house. The walls were cracked by these "Great Madrid Shakes" of 1811, Anderson's son, Charles wrote. "Its masonry must have been a gross fraud upon that very industrious and preoccupied employees."[6] In 1840, a bolt of lightning finished the house off, causing the walls to collapse. Fortunately the Andersons had moved to Chilicothe, Ohio in 1838 due to Kentucky's 1837 financial panic, when many banks failed.

Evidently, Locust Grove had no damage during this time. Major Croghan's knowledge of building was recognized and respected. In April 1788, he was ordered by the Jefferson County Court to lay out the new court house in the public square and to contract for its building.[7]

During 1960s, restoration of Locust Grove, Walter Macomber from Washington D.C. and Virginia was in charge of directing the project.

> I have stated before that the project at Locust Grove is one of the
> most remarkable with which I have ever been associated. I have

never known such coordination of purpose between city, county and citizens making such a worthy cause possible. In addition to this fine relationship Locust Grove will, I am sure, become one of the most outstanding buildings in Kentucky because of its fine architectural features. Its walnut paneling in its sophisticated design is outstanding and rare within the scope of my knowledge.

During the years of restoration of such buildings as Mount Vernon, Scotchtown—Patrick Henry's home, James Monroe's Law Office and others, I have never found such an exciting discovery as the 2nd floor "Ball Room" with its fragments of wall paper that covered the walls of that stately room.[8]

In *Restoration of Locust Grove*, author Sam Thomas wrote in detail how bricks were made on the property for building Locust Grove. "A primitive terra cotta water pipe system ran westerly from the well to an area where it appeared the brick had been puddled and baked."[9] At the time, one white male and seventeen blacks were living there, providing adequate labor.

When Major Croghan was building Locust Grove in 1790, he had plans for building his commercial trade on the Ohio and Mississippi Rivers. Because of their lengths and the tonnage these rivers could carry, they led all others in commercial trade. Before 1800, the farms along the rivers produced an enormous amount of farm products to be sold down the river in New Orleans. In some cases, both the boats and cargo were sold. Boats were outfitted with sails and sent out to sea when they reached the gulf.[10]

This is the option Major Croghan's brother-in-law, Colonel Anderson, chose. He equipped his boat with sails, and sent it out to sea where it immediately went under. This ended Anderson's adventures on the seas and his trade on the rivers. His son Charles wrote about his father and his financial affairs, "For neither he, nor either one of his sons, (except his two eldest sons) ever had the slightest pretensions to their name of being a real thorough or all-around business person."[11]

The 1807 Embargo Act enacted by the British prohibiting shipping to and from the Eastern Seaboard, meant financial gain for the westerners along the Ohio and Mississippi Rivers.

Wealthy country gentlemen ordered luxuries from the old country through agents, so their homes were filled with silver tableware, furniture, clothes, wines, and books. Acquiring these fine pieces was part of a circle of commerce that started with the slaves rolling the large hogsheads of produce down to the planter's wharf. This was then shipped out to merchants in New Orleans; then luxuries ordered from the old country were picked up and put on boats for the trip back up the river. Major Croghan's agents were the Gratz brothers in the east and his relative Daniel Clarke in New Orleans. Daniel Clarke made it possible for Major Croghan to ship produce to New Orleans by securing a special pass for him.[12]

These families lived like wealthy country gentlemen in Europe. Their country seats usually had large brick Georgian style homes looking out upon broad acres.

> In Louisville there is a circle, small tis true, but within whose magic round abounds every pleasure, that wealth, regulated by taste, can produce, or urbanity bestow. There, the "red heel" of Versailles may imagine himself in the emporium of fashion, and whilst leading a beauty through the maze of the dance, forget that he is in the wilds of America.[13]

Gorham Worth's testimony in Cincinnati was similar. "Talk of the back woods! said I to myself, after dining with Mr. Kilgour. I have never seen anything east of the mountains to be compared to the luxuries of that table! the costly dinner service,—the splendid cut glass,—the rich wines."[14]

Yet wealthy families, as well as the less affluent pioneer farmers, had heavy responsibilities. Most Kentuckians were still living on small farms or in rural areas. The ax and rifle were their most important possessions and the Indians their greatest enemies. Small farmers were most often isolated and lived in danger of attacks. But, if they endured the hard work and took advantage of opportunities offered on the frontier, their second and third generations could become wealthy land owners.

The frontier was not without its disreputable characters. The trade business down the river in New Orleans was the starting place for James Wilkinson's plan for the Spanish Conspiracy. The traitor was playing both sides, pretending to be working for Kentucky and at the same time being paid by the Spanish in New Orleans for bringing secrets of the frontiersmen's plans to open the shipping port.[15] This is where the famous Aaron Burr conspiracy came into fruition. Wilkinson was given charge of granting special privileges to navigation of the Mississippi, but all had to be in secrecy.

Spain had closed the shipping port in New Orleans to the farmers along the Ohio and Mississippi Rivers soon after the Revolutionary War ended. Kentuckians were desperate for action to open the Mississippi River to export their produce. They also were pushed to anger when Congress failed to grant statehood after three conventions had been held for petitioning Congress. The people alone would not need help of Congress whatever, if they decided to dislodge every garrison the Spanish had "with ease and certainty"and General George Rogers Clark would lead them.

France was preparing to move against Spanish Louisiana at the same time as General Clark and soldiers. The two factions with the same pursuit came together. France would blockade the Mississippi River by sea and General Clark would enter New Orleans from the Mississippi. Even though the General was so bitter against the government for the disgraceful treatment he had received, he had no desire to instigate a war between the United States and Spain. This is where he made his mistake by suggesting to the followers, "We must first expatriate ourselves and become citizens of France. My country has proved notoriously ungrateful of my services."[16]

This finally caused Eastern powers to take notice of those on the frontier who had been asking for help. Kentucky Governor Isaac Shelby was being pressured to put a stop to this planned expedition, but he sent a letter to Washington which was "calculated rather to increase than to diminish the apprehensions of the General Government as to the Western country. This letter had the effect desired, it drew from the Secretary of State information in relation to the navigation of the Mississippi, and satisfied us that the General Government was, in good faith, pursuing the object of first importance to the people of Kentucky."[17]

Unaware of Wilkinson's goal to destroy others and be ready to profit from their losses, General Clark's name was blemished by his exchange of letters with Wilkinson. When Wilkinson's conspiracy was exposed, Thomas Jefferson had just returned from Europe at that time and listened to Wilkinson's lies. Jefferson was convinced that everyone else, especially General George Rogers Clark were traitors to the country, except Wilkinson, himself. The damage was done when Wilkinson then went to Washington and spread scandalous reports about General Clark, Edward Livingston, and others who were innocent, paid a high price for being drawn into Wilkinson's scheme, also.

The shipping was opened once again by those in Washington, but for years, when General Clark's name was mentioned, the public would think of one image, the Spanish Conspiracy. It was said, this subject would throw the General into a rage—go to his room—drink and would not come out until sober.[18]

Wilkinson was on the scene when Lewis and Clark were held up in St. Louis during the winter of 1803–1804. They were waiting hourly for orders to arrive from New Orleans confirming that the Spanish Governor was to give possession of the territory of the Louisiana Purchase.

A Spanish decision to intercept Lewis and Clark came precisely at a time when Wilkinson, now stationed in New Orleans, was conniving with officials for profit and had sent guidance to the Spanish to "Intercept Captain Lewis and party, who are on the Missouri River, force them to retire or take them prisoners." Lewis and Clark, unaware they were being hunted, were always at a safe distance and out of danger from those in pursuit, who had not planned and prepared for the hardships of traveling the unknown territory and very soon fell by the way-side.[19]

Some events in General George Rogers Clark's life could be considered blessings in disguise. He lived his last nine years at Locust Grove surrounded with love and attention. In Lucy's home, the General was given a bedroom on the first floor. A narrow stairway from Clark's bedroom led to Kitt's room below. It was reported that Clark would tap on the floor with his cane to summon Kitt when needed. Conditions improved for Clark since moving in with the Croghans. Each day he sat next to Major Croghan at the end of the dining room table. The Major did not keep drink completely away from his old friend, but he mixed the General only one toddy each day at twelve o'clock.[20] Major Croghan handled Clark's correspondence, settling accounts and even paying some of his debts. He took care of the old hero.

Very soon after he came to live at Locust Grove, Major Croghan was instrumental in acquiring a pension for General Clark. The Virginia assembly finally voted to award the hero a pension of $400 yearly and a new sword, which was manufactured for him with elaborate engravings. Major Croghan kept records which show payments were made for Clark's attorney, notary public, physician, taxes, postage, and $310 for a new carriage and harness for him to attend a 4th of July celebration in honor of the old hero General Clark.[21] A new suit of clothes was ordered for this celebration and another was made for the presentation of the new sword, which took place in the parlor at Locust Grove in 1813. From merchants Fitzhugh and Gwathmey, a tailor provided clothes for both George Rogers and Kitt.[22]

John James Audubon who became famous for his paintings of birds wrote about a Fourth of July celebration which took place in 1811. "I knew a gentleman who had a large mill by the Falls of the Ohio. I became acquainted with the amiable Major William Croghan.

> Beargrass Creek, one of Kentucky's many beautiful streams, meanders through a deeply shaded growth of majestic beech woods, among which are walnut, oak, elm, ash and other trees. The spot where I saw an anniversary celebration of the glorious Independence Day is on its banks, near Louisville. The dense woods spread towards the Ohio shores on the west, and the gently rising grounds to the south and east. Every clearing holds a plantation that smiles luxuriance at the summer harvest. The farmer can admire both his orchards that seem to bow low to mother earth with fruit and his leisurely grazing flocks.

> The free, single-hearted Kentuckians, bold, erect and proud of their Virginia heritage had, as usual, made their plans for this occasion. The whole neighborhood joined in, with no need of an invitation where everyone—from the Governor to the ploughmen were welcome. All met with light hearts and merry faces. It was a beautiful day of blue heavens. Gentle breezes wafted the scent of gorgeous flowers. Little birds sang their sweetest in the woods, and the fluttering insects danced in the sunbeams. Columbia's sons and daughters seemed to have grown younger that morning.

> The undergrowth had been cut away, and the low boughs of the trees lopped off. A carpet of green grass formed a clearing that was like a sylvan pavilion. Wagons moved slowly along, bearing provisions from the farms hams, benison, an ox, and turkeys and other fowl. Flagons of every kind of beverage were to be seen. La Belle Riviere, the Ohio, had provided.[23]

General Clark had many true friends and one among them was Francis Vigo in Vincennes. None was more loyal to him. Two weeks later after the celebration took place, the General received a letter:

Vincennes July 15, 1811

Permit an old man who has witnessed your exertions in behalf of your country in its revolutionary struggles to address you at the present moment. When viewing the events which have succeeded those important times, I often thought that I had reasons to lament that the meritorious services of the best patriots of those days were too easily forgotten & almost taxed my adopted Country with ingratitude. But when I saw that on a late occasion, on the fourth of July last, the Citizens of Jefferson County and vicinity, from a spontaneous impulse of gratitude and esteem had paid an unfeigned tribute to the Veteran to whose skill and valor America and Kentucky owe so much, I then repelled the unwelcome idea of national ingratitude and my sentiments chimed in unison with those of the worthy Citizens of Kentucky towards the savior of this once distressed Country. Deprived of the pleasure of a personal attendance on that day, I took this method of manifesting to you, Sir that I participated in the general sentiments.

Please Sir to accept this plain but genuine offering from a man whom you honored once with your friendship, and who will never cease to put up prayers to heaven that the evening of your days may be serene and happy.

I have the honour to be Sir Your Most Odet[obedient] Sert. Vigo.[24]

For years there have been questions and controversies about three swords presented to George Rogers Clark by the state of Virginia. The first sword was broken by an enraged Clark and thrown into the Wabash River when after the capture of Vincennes, Virginia refused to pay Clark or his soldiers. He received the refusal to pay and the sword, one and at the same time while waiting for supplies to feed his soldiers. Clark, "...took the fine sword, walked out on the bank of the river ... thrust the blade deep into the ground, & gave the hilt a kick with his foot, broke it off & sent it into the river, & sent word to the Governor of Virginia, that he would have no such hollow-heated insignias while they refused his starving soldiers the common necessaries of life."[25]

CHAPTER NINE

War of 1812

No preparations had been made for going to war in 1812. The few men who made up the armed forces were not trained or prepared and certainly had no available supplies. Being unprepared for war, the fall of Detroit left Americans little to be proud of until they heard about a young hero and a place called Fort Stephenson.

After Americans had lost Detroit, the British General said to Tecumseh, "Show me what kind of country we have to march through."[1] Tecumseh drew a map on a roll of elm-bark and the march began.

The news of a brave and heroic American victory reached villages, towns, and cities in the Fall of 1813. A twenty-one year old Kentucky boy, Colonel George Croghan, with one hundred fifty men had beaten the British. The cry throughout the nation and in newspaper headlines was, "Hurrah for Croghan! Croghan! Croghan!"[2]

Under what circumstances, did Lucy Croghan hear the news of her son George and his heroic deeds? She was well aware of the notoriety that came with being a sister of two national heroes, but where was she and how did she learn about her boy?

When historians sought information about the Clark and Croghan families, they were referred to Eva Emery Dye, the recognized family historian. One hundred years ago, Eva Dye traveled east and for many months visited descendants for gathering family stories and copying family letters. William Clark's grandson, William Hancock Clark, made contact with relatives and set up appointments for Dye's visits in Montana, Washington D.C., New York, Virginia, St. Louis and many other places. Eva Dye copied many letters and wrote about family happenings as they were told to her.

The following gives an account of what happened when the news reached Louisville, Kentucky and Lucy learned of her son's brave defense of Fort Stephenson:

At Louisville, Kentucky, two mothers, Lucy and Fanny, were anxious for their boys. Both George Croghan and John O'Fallon had been with Harrison at Tippecanoe. Both had been promoted. Then came the call for swords. Both sons had made a call for swords to the folks back home.

Lucy had just read a letter from her fiery, ambitious son, George. She knew George was in an obscure fort on Sandusky River near Lake Erie. George wrote, "The General little knows me. To assist his cause, to promote in any way his welfare, I would bravely sacrifice my best and fondest hopes. I am resolved on quitting the army as soon as I am relieved of the command of this post" The two mothers were sitting under the trees at Locust Grove. Lucy and Fanny heard a crowd yelling in the distance coming down the road from Louisville. As they drew near, they were yelling.[3]

Hurrah for Croghan! Croghan! Croghan!

Lucy went to the gate and heard the story from several lips at once and in hurried words, "Why, you see, Madam, General Harrison was afraid Tecumseh would make a flank attack on Fort Stephenson, in charge of George Croghan, and so ordered him to abandon and burn it. But no, he sent the General word, 'We are determined to hold this place, and by heaven we will!' "[4]

"Hurrah for Croghan!" was shouted again and again down the streets of Louisville and the bells rang out a peal as the Stars and Stripes ran up the flag-staff.

"The little game cock, he shall have my sword," was supposed to have been said by George Rogers Clark, who once again was living his own great days.

Louisville wasn't the only place ringing the bells of celebration. The whole nation was running the Stars and Stripes up the flag pole. "Then, word began to pass that this young twenty one year old hero was a nephew of another hero who had held these red men at bay without provisions, no shoes and almost no army, 'and does he yet live?' " It was soon discovered George Rogers Clark lived on the shore of the Ohio River. "Has he no recognition?" they asked.

General Clark's nephew's brave defense at Fort Stephenson touched the old war hero like nothing else had and he repeated, "Yes, yes, he shall have my sword!"

George Croghan with 160 men and one six-pounder fought off British General Proctor with 500 regulars and 700 to 800 Indians. Under Croghan's planning and directions, a ditch nine feet wide had been dug around the fort in preparation for an onslaught. Fire was held until the enemy filled the ditch, then the British soldiers and Indians were falling and trying to escape. This lasted almost until morning when the boats left and returned to Canada.

This young soldier was the most celebrated in the entire land. Newspapers across the nation couldn't praise young George Croghan enough for his heroic deeds. This standoff at Fort Stephenson added more fame to Locust Grove. The Croghan name was already well known in the country, due to Major Croghan's many important acquaintances.

There was little normalcy in Lucy's daily household routine at that time. This old home place was a second home for William Clark and his family, especially after moving to St. Louis and making frequent travels to Virginia.

Daily routine was interrupted by many visitors who came to see General Clark, including Indian delegations. In 1817, a delegation of Indian chiefs paid a formal call on General Clark at Locust Grove and found him seated on the long north porch.

> The son George was married in 1817, so it was later than that the ceremonial Indian wooden spoon (now in the Locust Grove Museum) was used. All those present at Locust Grove were seated in a large circle on the back lawn, taking part in a ceremony. My Great Grandmother, a New York Livingston, was obliged to take her turn eating (I believe it was fish) when it came her turn, her husband whispered to her that she must eat or the Indians would be greatly offended if she did not.[5]

This visit paid to the old general was not uncommon. Years later, 1847, nephew John O'Fallon, living in St. Louis remembered the Indians visiting Mulberry Hill and afterward at Point of Rock.[6]

At Croghan's country seat, Lucy was surrounded by people at all times. Her attention must have been pulled in many different directions at the same time. There were times when nephews made their home with their Aunt and Uncle Croghan at Locust Grove.

July 29, 1815, nephew R. C. Anderson, Jr. wrote in his dairy that he went to the Quarter races below town and wife Betsy went home with "Aunt Croghan" the day before. The next day he arrived at Locust Grove and wrote in his diary, "Genl. GR Clark told me that he never knew an Indian to have the toothache."[7] This seems to be a reliable source for the claim that General Clark's speech was not impaired at this time.

While Lucy's brother Jonathan was living, he was an overnight guest at least three or four times a month and his diary shows three or four days each time. Some of his sick stays lasted ten days while at Locust Grove. Later years, Lucy nursed her sister Fannie through periods of sickness. Lucy's sister, Ann Clark Gwathmey, died at Locust Grove. She wouldn't have found much free time when her brother William, Julia, and children visited. Lucy must have been a strong woman, both physically and mentally, to have withstood all the care she gave so many.

There were some things in life that Lucy's care couldn't fix. In September, 1814, her young brother William wrote from St. Louis, "I think it not improbable that Mrs. Clark will have to take another trip to Phila. Her breast continues to be sore & enlargens. She has been at the Sulpher Srpings for six weeks without any Provceible change for the better. I am apprehensive it will terminate in a Cancer."[8]

It was at Lucy's home in February 1818 that another historical event took place. General George Rogers Clark, the old soldier, died.

An Old Soldier Dies

"The mighty oak of the forest has fallen,
and now the scrub oaks may sprout all around."
Judge John Rowan
The Ky Reporter of Feb 25, 1818

The old General who captured British officer Hamilton
during the Revolutionary War
Died at
Locust Grove on 13th of February 1818.

After hearing the news of the old General's death, the court of chancery immediately adjourned for the day. Members of the bar, having convened, resolved to attend the interment of General Clark. They resolved to attend as a group. The members of the bar, as a testimony of their respect for the memory of General Clark, wore crape on the left arm for thirty days.

'It was a stormy, snowy day,' when the funeral was held at Locust Grove on Sunday, February 15, notwithstanding the disagreeableness of the day of his internment, the crowd that assembled to pay this last tribute to his remains, was truly great. It was a source of melancholy, gratification, to see mingling with the crowd, a few of his old revolutionary associates, who also wore a band of black crepe on the left arm.

The Rev. W.C. Banks officated in his professional capacity, by offering up an appropriate prayer to the throne of grace and was succeeded by the Hon. J. Rowan gave an impressive eulogy on the character of the ever memorable hero. Every eye was suffused in tears, elicited by the sympathies of the heart, which were made sensible of the debt of gratitude they owed their deceased benefactor. The peal of artillery announced the commencement of the procession which was to escort the remains of the revolutionary warrior to his last abode. Minute guns were fired during the ceremony, and until the ground was raised upon that form which was once the shield of his country and the terror of her foes.[9]

CHAPTER TEN

A Day in the Life of Lucy

During Lucy's lifetime she was truly a caregiver. Where did she go for quiet times and rest to collect her thoughts and to reflect on her life? Was it in her garden? At that time a garden was considered an outdoor room, used in fine weather for entertaining. Climbing arbor vines provided shaded alleys for walking and resting. Old gardens had what they called "surprises," places to sit and rest. Soon after visitors arrived, and shown their rooms, a stroll through the beautiful landscaped garden was planned for show. A common practice and custom of the day was the exchange of a new or rare flower or plant.

Leisurely strolls walking down the winding paths in the late evening, when the most fragrant scents are heavy from the flowers, provided a romantic interlude in the lives of young people, even then under watchful eyes not too far away. Some "addresses" (marriage proposals) took place on these walks. Men discussed business while strolling through the garden.

Morning walks in a garden were close to a spiritual experience for Lucy, with the singing of the birds and butterflies going about their business. The garden would also have been a place of refuge when her heart was heavy. At that time, William's wife, Julia, was ill and growing weaker.

Cutting flowers for arrangements in the house was part of the daily routine. Those old-fashioned flowers, especially roses, had a fragrance that permeated the room, which would tend to neutralize some of the barnyard smells always in the air as well as having an effect of a welcoming beauty for those who entered. Climbing roses were planted on the back side of the house and soon climbed to the second floor to counteract the unpleasant odor from the service yard.

Major Croghan would have landscaped gardens and groves of trees planted at Locust Grove. A Gentleman's Country Seat was described as

Historic Locust Grove, photograph by John Nation (from the collection of Historic Locust Grove, Inc.)

"having vast trimmed lawns, tree groves, vistas, and formal gardens which presented an extraordinary view to the numerous guests." In family letters, Locust Grove is referred to as the "grand old mansion."[1]

A road had to be cut from River Road to make rough passage for coaches and carriages through the orchards. Major Croghan named this Stanwix Road for Fort Stanwix, near Sir William Johnson's home in New York. The name Stanwix was later changed to Blankenbaker Lane and the road itself was later changed and brought closer to the Mansion House in 1963.[2]

Lucy Croghan's nephew, J. Wyatt Jones (Ann Clark Gwathmey's grandson) practically lived with his Croghan cousins, Charles, Nicholas and Edmund during the summer months when they were growing up. Years later while living in St. Louis, he wrote about his memories with the Croghans at Locust Grove "commanding an Extensive View of the Ohio River."[3]

Major Croghan's way of doing things was in grand style. When President Monroe, General Jackson, and their retinue were weekend guests in 1819, he would have offered no less than a cultivated and landscaped vista for greeting the honored guests in their four horse drawn coach.

Leading up the hill through the huge acres of orchards, made a beautiful and impressive entrance "for show." When Locust Grove came in view, "To the left of the Back door was the carriage house and the Shop where the forge and anvil were used. Beyond this was the apple and peach orchard across the road from this orchard was another orchard of cherries and peaches."[4] These carriage houses or coach houses were handsome buildings which complimented the main house in appearance. The second floor was used for living quarters, most likely for young visiting gentlemen. Tax records for 1816 show that Major Croghan had four (fourwheel) carriages at that time.[5]

Approaching the front iron gateway entrance with bricked piers on each side, there appeared a spacious lawn with a circular drive which still exists two hundred years later. The view from the front entrance would not have been obstructed by fences. The ha-ha wall was built for that purpose.

From Alfred Beckley's Memories of Locust Grove in 1809–1814:

> In the year 1814 (being in my twelfth year) my mother broke up housekeeping and took me with her on a second journey to Kentucky. During our stay in Louisville we visited "Locust Grove," the seat of Major William Croghan, the father of Major George Croghan, the Hero of Sandusky, and my mother's dear friend Dr. Croghan, where he dispensed to all visitors Old Virginia and Kentucky Hospitality. The ladies had started to the Grove in Carriages and Dr Croghan hastily started me upon an old gentle superannuated horse, telling me to follow the river road some five or six miles and to turn off to the right at an old cabin which would take me to the Grove. Somehow I overlooked the old cabin and went on till I came to a large creek very high from backwater from the river and past fording. I hesitated and as the Doctor had said nothing about a creek, I turned back and very providentially as I was afterwards told that my old horse could not swim. On reaching the cabin a black woman gave me directions to reach the Grove, where I found my dear mother and the whole family quite uneasy about me. While there, we saw the illustrious Indian fighter, the Hero of Kaskaskia and Vincinnes, Col. George Rogers Clark. He was a paralytic, his mind pretty much gone. He was drawn about the house in a chair upon wheels.
>
> While at dinner a Negro boy engaged in pulling a series of small flags suspended from the ceiling to keep the flies off the table was pointed out as soundly asleep and yet mechanically pulling the cord and keeping the flags flapping. To convince the company, he

was suddenly awakened and proved there was no deception in the matter. It produced much laughter and merriment.[6]

With so much coming and going at Locust Grove, the pattern of daily housekeeping was altered from day to day and from season to season. In the fall when heavy field work was ending, Lucy surveyed her stores of warm clothing and bedding for the coming winter. Cellars, attics and pantry shelves would have been crowded with foodstuffs of all kinds. Increased artificial lighting would be needed.

Being prepared for winter took careful planning and management. Roaring fires in January and February could not keep the far corners of a room warm. Doors between rooms were kept tightly closed and people would hurry through icy halls and passageways. Brass or copper bed warming pans filled with red hot coals from the kitchen or sitting room fireplace would be taken to the cold bed chambers for temporary comfort and to reduce some of the dampness. Water in washbasins would freeze during the night.[7]

Winter evenings were long and most families gathered in one room. This offered opportunities for family activities such as board games and music, and for reading, knitting, correspondence, and writing diaries. A very important time was that spent for moral instruction.

The young ones heard many stories of the first glimpse of land from a ship, the ring of an ax in the woods, an Indian war-hoop, celebrations of freedom and at that time in history, Napoleon's war in Europe. Youngsters also, learned about the family members who had given so much while building the new nation. For sure, the story of Uncle Jonathan Clark's presence in the little church at Richmond and the Clark brother John, who died of tuberculosis as result of being a prisoner on a British ship. William Clark was only six years of age when he saw five of his brothers go to war. He hated the British with a passion for the rest of his adult life.[8]

There were many Revolutionary heroes in the Clark family, including Major Croghan himself, who fought the British, but it was highly unlikely American patriots, especially in the Clark family, would want to be reminded of someone's British relatives after the war ended. One young lady wrote later, "My Father wouldn't allow we children to play with 'Tory' children, 'nay, even speak to them.'"[9] Thus, the Croghan children obviously didn't hear much about Major Croghan's British relatives who fought for the enemy, but in later years, as Dr. John Croghan viewed the Ireland coast from a ship, he was flooded with memories of the stories his father had told about his homeland.[10]

Spring would arrive and activity would come alive with cleaning. Boots were often muddy, and so many tramping endlessly over the floors would create problems for the mistress of the household. To clean the floors, damp sand was scrubbed across the floor, allowed to dry and at the same time walking on the floors. Then the floors were swept clean, leaving a fine patina.[11] Dolly Madison solved the problem while living in the White House. In the entrance hall, boots and heavy shoes were pulled off, replacing them were thin slippers which were clean and kinder to the canvas and carpet on the floors.[12]

In warm summer days, the central hall at Locust Grove would have been used as a sitting room. When Locust Grove was restored as a historic home and opened to the public, docents sat in the downstairs center hall with front and back doors open, allowing cool breezes to pass through the house, just as Croghans had done so many years before.

In the spring and summer when the gardens and orchards were full of their bounty, each morning the cooks who worked in the outside kitchen would sit beside the kitchen wall [outside] preparing vegetables for cooking. The brick walk which ran along the kitchen was not level and sloped to the center, forming a trench. A wooden trough with one end propped upon the edge of the well and the other end in the trench, was used to pour water from the well into the trench. The cooks could dip their fresh vegetables down into the water, cleaning them for cooking. During the restoration years, Walter Macomber, who was in charge of restoration told one of the ladies to leave the trench as it was, explaining the purpose for the narrow trench in the brick walk.[13] However, in recent years, the walk at Locust Grove has been covered over with a boardwalk, making smoother walking for guests on tours.

Young girls would start training very early to run a household and manage servants. They learned to read, and do arithmetic well enough to keep household accounts and read the Bible, but they also had to learn to make a ball gown, as well as a slave garment. When they became mistress of their own homes, they would then have to train one or more slave women in this skill in order to turn the sewing over to them.

Girls also, made samplers to learn their needlework stitches. Usually the name of the girl who did the sewing, as well as her age, the date the sampler was made, the letters of the alphabet and numbers from one to ten were cross-stitched on the samplers. Mottos were particularly popular as sampler themes. A sampler made by one of Jonathan Clark's descendants is framed and displayed on the second floor at Locust Grove.

During the 18th and 19th centuries, sewing elevated from a craft of necessity to a creative art and parlor activity. Elegant wedding gowns were made and petticoats were embroidered with exquisite detailed work and fine stitching.

Quilts made by Lucy's nieces, the Anderson girls, are in the collection at Colonial Dames Dumbarton House in Washington, D.C.[14] The Croghan girls would have made "company" or a "bride's quilt" for show.

Needlework was a freedom of expression for the mistress who couldn't vote and wasn't consulted about what went on at the slave markets. No one could tell her what design or colors she would choose for her quilt. Needle work could be therapeutic, especially when the need for a quiet place arose.

From: *Aunt Jane of Kentucky*

> It took me more than twenty years, nearly twenty-five, I reckon, in the evenings after supper when the children were all put to bed. My whole life is in that quilt. It scares me sometimes when I look at it. All my joys and all my sorrows are stitched into those little pieces. When I was proud of the boys and when I was downright provoked and angry with them. When the girls annoyed me or when they gave me a warm feeling around my heart. And John too, he was stitched into that quilt and all the thirty years we were married. Sometimes I loved him and sometimes I sat there hating him as I pieced the patches together. So they are all in that quilt, my hopes and fears, my joys and sorrows, my loves and hates. I tremble sometimes when I remember what that quilt knows about me.[15]

It seems that dancing school during the summer was high on the list of activities for young people. Jonathan's Diary reads: "Monday July 1805— 3, Clear at Majr. Croghans, Dancing School." The diary also lists "Thursday, August 1805—1 Clear Dancing School 2nd day at my house." The next, "June 1806 Dancing school had begun."

Dancing skills and music appreciation were expected of well brought up people. Ann Croghan played the piano and William Croghan, Jr. played the violin.[16] All the children at Locust Grove had music instructors and dance masters brought in for lessons which took place in the ballroom. They had to master "the accomplishments" of dancing the intricate steps of the Allemande, the Gretagne, and the Rigadoon, Minuets and French dances were especially popular. Sometimes, the country dances, such as the Virginia reel were danced. Dolly Madison liked these to add a little flavor.[17] In Boston, 1806, the old-fashioned minuet was still being danced, although the waltz was considered not respectable.

When Clark cousins gathered at Locust Grove for dance lessons, they could have danced to music from a music box. A music box is listed in the long inventory of furnishings when the estate settlement of Locust Grove was probated in 1849.[18]

Singing school was also important in their lives. Richard C. Anderson, Jr., wrote in his diary, "September 16, 1818 . . . My wife [Betsy Gwathmey] has been taking lessons at the 'Singing School' kept at my father's [Soldiers Retreat] by Metcalfe."[19] This form of singing still exists in the south and is called "old book" or "shaped note" singing, and consists of beautiful vocal singing. This form of singing came over to America when the first colonists arrived, and through the years, was preserved as it moved down the Eastern Seaboard, through the Appalachian country.[20] Today, an annual event of the shaped note singing convention takes place in Benton, Kentucky and participants from the academic world of Boston, Chicago and universities in the South are represented.

Obviously, besides music and dancing, young women and men were educated very differently. The following was written by Thomas Jefferson to a schoolmaster, who had sought Jefferson's advice. Jefferson answered that for his daughters he provided:

> A solid education, which might enable them, when they became mothers, to educate their own daughters, and even to direct the course for sons, should their fathers be lost ... the time lost on novels, poison infects the mind, it destroys its tone and revolts it against wholesome reading.

> The ornaments, too, and the amusements of life, are entitled to their portion of attention. These, for a female, are dancing, drawing, and music. The first is a healthy exercise, elegant, and very attractive for young people. Every affectionate parent would be pleased to see his daughter qualified to participate with her companions and without awkwardness at least, in the circles of festivity,

of which she occasionally becomes a part. It is a necessary accomplishment, therefore, although of short use; for the French rule is wise, that no lady dances after marriage. This is founded in solid physical reasons, gestation and nursing leaving little time to a married lady when this exercise can be either safe or innocent. Drawing is thought less of in this country than in Europe. Music is invaluable where a person has an ear. It furnishes a delightful recreation for the hours of respite from the cares of the day, and lasts us through life.[21]

Abigail Adams had her own convictions about what a young girl should know and learn in preparation for her life in the future. Abigail upon reading her husband's letters of how the patriots at the Continental Congress in Philadelphia were agonizing over the Bill of Rights, was prompted to write the famous words, "I desire you would remember the ladies, and be more favorable to them than your ancestors. Do not put much unlimited power into the hands of husbands."[22] Little did the young women in Louisville heed Jefferson's advice. Charles Anderson wrote about his mother, "being the best dancer on the ballroom floor at Soldiers Retreat."[23]

Sadness entered William's life which would change his life forever. His wife Julia's health had declined so, Dr. Farrar suggested that a sea voyage and some time in the Virginia Mountains might be beneficial.[24]

Governor Clark with Julia, three sons, small daughter, and two maids came to Louisville for a visit. The Clarks then traveled to New Orleans before a voyage to Washington, D.C., which proved to be good for the children, in spite of the circumstances of Julia's health.[25]

In Washington, Julia rested in bed at the hotel, "receiving kind attentions from the ladies of the President's House." Clark was received by President Monroe and Mrs. Monroe in the Green Room. The next day, Mrs. Hay, the Monroes' daughter, then living with her parents, invited the boys to tea with her little daughter, Hortense Hay. Dolly Madison called on Julia with Mrs. Gallatin, and both were sure that Julia's upcoming visit to Sweet Springs would be of much benefit.[26]

But, the stay at Sweet Springs gave no improvement to Julia. She had to be carried to her parents' at Fotheringay on a bed. For three weeks, Julia was expected to die momentarily. Fumes of tar were administered through a tube and miraculously Julia's condition improved. The President had recommended William stay in Virginia for the winter, even though things were not going well for him back in Missouri. Clark had enemies in St. Louis and one of his nephews wrote, "A Major Hall has injured you in your finances,"[27] During William's last years as Governor of Missouri, some powerful citizens were very vocal about his failure to live up to their expectations when he was Governor.

In late spring, William Clark's sons, Lewis and William returned to school. Governor Clark returned to Missouri, but upon his arrival learned that Julia had passed away.

When Julia was dying, an old slave, "Granny Molly" who had been present at Julia's birth, once again was by her bedside. William Clark's daughter-in-law, the second wife of Meriwether Lewis Clark always remembered the family stories of Granny Molly. Little Lewis, then eleven years of age came to the bedside, with his curly hair disheveled, and his broad shirt collar tumbled, "Oh, watch over my Boy, and keep him neat, he is so beautiful, Granny", Julia said.[28] She died on June 27, 1820 and after her pretty wake in her parents' parlor, she was buried in the mausoleum that Colonel Hancock had cut from the solid rock on the hillside overlooking the green valley of the Roanoke.

Julia's brother George and his wife Eliza Croghan Hancock were at Fotheringay when Julia died. William hastened back to Virginia on the long sad journey. Only two weeks after Julia died, her father Colonel Hancock died. Some said he died over the loss of his daughter Julia.[29]

William had placed his little daughter Mary in an Eastern boarding school before he and his other motherless children started on their return down the Ohio River for their home in St. Louis. In his absence an election had unseated William Clark as governor.

When the family reached Louisville, they stopped at Locust Grove. William's brothers, Jonathan and George Clark were no longer living, and only his sisters Ann, Lucy and Fannie survived, but it would have been comforting for William to be with the Major and Lucy at Locust Grove.

Not long after the return to St. Louis, William received a letter from little Mary;

> Dear Papa
> I hope you are well. I want to see you, and my Brothers. kiss
> them for me. I am a good Girl, and will learn my book.[30]

Several months after Julia's death, Clark returned to Virginia to bring Mary back to St. Louis, but upon arrival in Louisville, Mary became seriously ill. William's brother-in-law Judge Fitshugh wrote to John O'Fallon, on October 16, 1821, "Gen. Clark has been at Mrs. Preston's [Julia's sister] for some time, attending to his Daughter, who is extremely ill, not expected to recover the doctor told me last evening she could not recover."[31] Little did he know that Mary had died during the night. Mary's death took place at her Aunt Caroline Hancock Preston's home, not far from Locust Grove.

Clark always kept the letter from Mary and wrote on it: "Miss Mary M. Clark's letter recd. 13th July 1821 answered by all the boys." William also lost his deformed son Julius when he was thirteen years of age.[32]

During these years, Locust Grove seemed to have become the heart's resting place for generations of Clarks and Croghans. One month later, after little Mary's death, William addressed a letter to N. Biddle from "Majr. Croghans Near Louisvlle, Kty. October 28th 1821."[33] William was anxious to learn of any prospects of gaining anything from the publication of the Lewis and Clark book.

The marriage of William Clark and Harriet Kennerly Radford, took place on November 28, 1821. This was "the other girl who he had helped with that balky horse so many years before."

After the death of Harriet's first husband John Radford in 1817, she and children, William, John, and Mary Radford had arrived in St. Louis to live with her brother, James Kennerly who had come from Fincastle, Virginia in 1813, at the invitation of Governor William Clark. Kennerly and Clark's nephew John O'Fallon had opened a store just before Governor Clark appointed Kennerly forwarding agent of St. Louis for transporting goods from the superintendent of Indian trade to the frontier forts and receiving furs in return from the Indians.

Stepson William Radford was very much against the marriage of his mother and it took some time for William Clark to win him over.[34] Two more children were born to the second Clark marriage, Jefferson Kearney Clark (1824–1900) and another son named Edmund Clark who died young. Harriet died on Christmas Day, 1831, after ten years of marriage and seven-year old Jefferson Kearney Clark went to live with his Uncle and Aunt Kennerly, who raised him as their own. It was said, that after Harriet's death, Governor Clark aged considerably over the next seven years, even though he continued to be busy arranging Indian treaties.[35]

CHAPTER ELEVEN

Social Life—Matchmaking—Marriage

Social graces, breeding and gentle manners were the order of the day. During the years, 1814–1824 at Locust Grove, there was laughter, partying, and romancing. William and Lucy Croghan's house was a social center for young people and Locust Grove's hospitality reached a height it had never met before and never would again. Matchmaking for new husbands and wives was not a small part of their lives since the Croghans and their many cousins and friends had reached the age for marriage. The three older Croghan boys were always bringing their schoolmates home for long visits and quite often these friends would fall in love with their Croghan sisters, Ann and Eliza. More than one young man asked for the hand in marriage of one of William and Lucy's daughters. Friends and cousins of the Croghan sisters also took part in the gaiety, and entertainment. This was when the privileged "Kentucky Aristocracy" came to full bloom.

Within these houses, such as Locust Grove, parties lasted for several days. In the winter time, there would be fires blazing in every fireplace and in the summer time, breezes would blow through the trees in the second floor ballroom windows. Cooling drinks would be served on the back porch and across the sloping lawns strolled the beaux and belles and through the gardens in moonlight among the roses, but always under watchful eyes.

The most eminent men of the city and country gathered in the library to discuss local politics and what was happening on the national scene. Major Croghan was civic-minded and active in the development of Louisville.[1] At that time there would have been talk about the banking system in Kentucky and discussions about building a canal in Louisville.

They came, cousins and friends, in hordes and they stayed for long periods of time. A schoolmate David Rozel Poignand, wrote about one of these visits.

> In December 1813 with a party of upwards of a dozen gentlemen
> and two ladies, mostly connections, I supped, slept and breakfasted

at Major W. Croghans, the brother-in-law of Geo. R. Clark. Several
of the party entered the room occupied by the General. The hum of
conversation and the laugh of the party was heard inside.[2]

Food and wine were plentiful and people were not fussy about expecting one bed to a room, or one person to a bed. The ballroom could have
been used as a dormitory for the young men, with pallets on the floor, and
the girls would have gotten the privacy of the bedrooms. Sometimes,
young unmarried men were domiciled in another building outside the
main house which most likely was living quarters above the coach house.
Indeed, they would have taken their baths in a room out back of the main
house, preferably with a fireplace during cold weather. They would have
also brought their own personal servants along to heat water or prepare
their baths and do their laundry, since fresh linens, such as shirts, were
important. The wash house would have been very busy at all times and
sometimes disputes would erupt between the servants.

The privies or "necessaries" would be strategically placed at the back
of the gardens for gentlemen.

When Locust Grove was bursting at the seams with young folks frol-
icking for days, sleep took place any time during the day or night, because
there was dancing till the wee hours of morning and many other activities going on in the daytime. Sideboards were laden with food and available at any hour of the day. It would have been very difficult to seat forty
or fifty people for a formal seating in the Croghan dining room, so more
than likely tables would have been set up in other areas downstairs.

There was plenty of time for games of billiards or cards, and for hunting. It was part of the sport to place the sideboard or "Gentleman's
Drinking Table" out under the trees with drinks. As the hunters would
return from their fox chase, they would ride by and try to pick up their
drinks without spilling any. Sometimes this sport was changed slightly to
"Racing for the Bottle." The sport of chase had no regard for danger to
the riders or horses through brush and muddy hollows and a shrill whoop
would be heard from the one in front as he approached the bottle,
although the bottle was shared with all the riders.[3]

There were parties in rooms, some cards and some singing of "Liberty
Songs," as they called them, in which six, eight or ten or more would put
their heads close together and roar.[4]

Young girls flirted coquettishly with their exquisitely designed folding fans and their kid gloved hands twirling parasols. The boys dressed like in European fashion and paraded around. Upon seating themselves, these male dandies would gracefully arrange their coattails. It was the habit of young gentlemen of this period to turn the armless chairs around, sit on them backwards and rest their arms on the tops of the chair backs.[5] The silver-tongued devils were spreading their charms.

Two Louisville historians wrote about the popular Gwathmey's pond and the skating scenes that took place each winter. Lucy's children would indeed, have participated in these skating parties. Gwathmey's pond was located within the present day boundary of Cedar, Walnut and Seventh Street. The land and house was owned by Lucy's nephew, John Gwathmey, the son of Owen and Ann Clark Gwathmey [Lucy's sister].

> In the winter, the frozen ponds and lakes provided a scene of many a merry party. On the moonlight evenings, numbers of ladies and gentlemen were to be seen skimming over its surface, the gentlemen on skates and the ladies in chairs, the backs of which were laid upon the ice and the chairs fastened by ropes to the waist of the skaters and thus they dashed along at furious speed over the glassy surface. Beaux and belles, with loud voices and ringing laughter and the merriment of the occasion was only increased when some dashing fellow, in his endeavors to surpass in agility and daring all his compeers, fell prostrate to the ice, or broke through into the water beneath.
>
> There was much color and gaiety around this lake, for there much of the social life of the city centered. Picknicking and pleasure boating were indulged in the summer months; bonfires blazed on it's bank in the winter time and Negro musicians fiddled for the delectation of ice skaters. There, too, baptisms were a regular occurrence; sinners went under water.
>
> After the skating party came to an end, the young people would go into a warm house for something hot to drink then gather around the piano for singing. Collection of songs were bought in song books.[6]

Some of the titles were: *"The Forget Me Not Songsters"* and *"Museum Of Mirth."*[7]

A Gwathmey nephew of Lucy wrote about his Uncle William and Aunt Lucy:

They resided at 'Locust Grove', the noted homestead a few miles above the city boundary, on and commanding an extensive view of the Ohio River. Mrs. Croghan was the only one of the daughters I remember and recall with pleasant many happy hours, I have passed at the old homestead.

Mrs. Croghan was a veritable Clark. She had that family pride, but with it, no austerity that would detract from the loveliness of her character. Being the sole surviving sister, she was venerated by her family connection and for her unostentatious hospitality and generality held in the highest esteem by all who knew her. My recollection of George Croghan coincides with your description of him. He was very much like his mother and the resemblance between him and his Uncle George Rogers Clark, judging from the portrait we have of him.[8]

Clark nephews and nieces, the Andersons, Gwathmeys, O'Fallons, Thrustons, Fitshughs, Clarks, Croghans and their neighbors, such as Zachary Taylor and siblings, seemed to be together constantly. All of this ended as cousins and friends had begun to choose their mates for marriage. Those who had not moved elsewhere joined in the happiness and celebrations as each entered into marriage.

It was the summer of 1817, and one can imagine Major William Croghan and Lucy glowing with anticipation of having all their children together, once again, after years of absence. The Major wrote to cousins, aunts, uncles, and friends, inviting them to join in their happiness. His brother-in-law William Clark (St. Louis) wrote his regrets to William Croghan, Sr., on May 18, 1817, "We would be much gratified in participating in the happiness you each enjoy on the return of all your sons, but I fear it will not be in our power to visit you very soon."[9]

The Major and Lucy were about to welcome a new daughter-in-law into their family. George, their second son, was the first to marry on May 15, 1817. At the height of fame he met Serena Livingston, a New York society lady from the Livingston Hudson River family. After the wedding, Colonel George brought his bride to his home in Kentucky where his family and friends were waiting to meet the niece of General Montgomery, General Armstrong, and Chancellor Robert Livingston from New York.

Southern hospitality would have called for a social event with all the Clark kin present. Cousin Richard C. Anderson, Lucy's nephew made note of this event in his diary.

> June 24, 1817. I was in Louisville. Colo[nel George] Croghan &
> Lady from N York arrived at Majr.[William] Croghan's.[10]

Lucy's two daughters were the next to marry and William Croghan, Jr. was the last. Only four of the nine children married and only George, William, and Ann would become parents of the new generation. In 1819, Louisville came alive and the "old guard" turned out to welcome a visit from President Monroe and General Jackson. Major Croghan presided at the banquet given for The President of the United States and the General. *Croghans March* was played at the banquet. The march had probably been written for the Major's son George Croghan. The entire visiting party stayed at Locust Grove for the weekend.[11] Locust Grove would have been a beautiful place in June, with everything freshly in bloom.

> Louisville, KY. June 26, 1819
>
> On Wednesday last, the President of the United States and suite accompanied by General Jackson and his suite arrived in this town. The following day, being the anniversary of St. John The Baptist and Masonic Oration, was delivered at the Presbyterian Meeting House, by William Croghan, Jr. and afterwards a Masonic dinner was given then at the Union Hall Tavern and a Ball on the same evening at the Washington Hall. On this occasion, as might be expected, every respectful attention has been paid, and no doubt many a pleasing recollections recalled by the President and his old acquaintances, some of whom were his play fellows in boyhood, and others, companions in arms during the Revolution, are not now to be found in so small a space as in this town and neighborhood.
>
> On this day, the parties will leave here for Jeffersonville, IN, where a public dinner will be given, from whence they proceed to Major Croghan's for the night.[12]

When a young man is about to get a wife, the first inquiry he makes is, "has such a young lady much property?"

The future sons-in-law of Major Croghan and Lucy, Thomas Jesup and George Hancock, more than likely were part of the President and General's retinue, since both were in Louisville at that time. Thomas Jesup had been a protégé of President Monroe since 1814. George Hancock, son of a former U.S. Representative and one of the largest land owners in Virginia, had relatives in Louisville. George Hancock and William Croghan, Jr., had been classmates at Carlisle, Pennsylvania.[13]

The Hancocks were related to two presidents, James Madison and Zachary Taylor. George, Julia and Carolyn Hancock's father was in the Virginia General Assembly in 1784 and served for eight years before being elected to the U.S. House of Representatives. During this time, the Hancock family lived in Philadelphia, traveled in a four-horse carriage as "colonial gentry."[14]

Two other prominent families, the Livingstons and Lees lived in Philadelphia at the same time the Hancocks were there. Those three families became very close in Philadelphia while the new nation was being formed. Family letters reveal that Serena Livingston was in love with Henry Lee who was the older half-brother of Robert E. Lee, but later financial misfortune had fallen on Henry Lee and was under much stress since his father had gone to debtor's prison. Henry married a neighbor girl whose parents had died, leaving a wealthy daughter with an estate. George Hancock was in love with Ann Lee, Robert E. Lee's sister, who married William L. Marshall of Baltimore.[15]

Hancock had visited his sister Caroline Preston and his Taylor cousins in Kentucky before 1819 and wrote about seeing General Clark at Locust Grove. The Summer of 1819 when President Monroe visited Locust Grove, Eliza was eighteen years old and was a real beauty. George Hancock had attended Carlisle College at the same time as William Croghan. A classmate who became President of Carlisle College said that George was the most handsomest man he had ever seen.[16] Eliza became George Hancock's bride on September 28, 1819 at Locust Grove.[17]

George Hancock's father was not only one of the largest landowners, but in Virginia he was co-owner of the Alleghany Turnpike. He was a successful lawyer and politician, besides managing his rural estates. One of these was a magnificent showplace manor house, Fotheringay, which George inherited. Eliza went to live there as a bride. One year later, she was mistress at Fotheringay.

George's mother continued to live at Fotheringay after turning the mansion over to George, even though she had a life claim on the home. Ann Jesup came to stay with her sister Eliza at Fotheringay when Ann's second child was born, in September 1824.

Eliza Croghan Hancock, artist unknown (From the collection of Historic Locust Grove, Inc.)

Thomas Jesup had moved up in society after leaving his roots in Western Kentucky to join the army. He had arrived in Kentucky with his widowed mother and young brothers and sisters at a very young age. At nine years of age, it was told, he was working to help support his mother and siblings. It was a long journey from this beginning to being an aide-de-camp to President Monroe, who had relied on Jesup as a trusted favorite at a very young age.[18] Jesup's return to Kentucky for the visit at Locust Grove in 1819 could have been the first meeting with Ann Croghan.

Colonel George Croghan and Thomas Jesup had been friends during the War of 1812. So Jesup, as well as Hancock, was very much aware of the wealth and land properties owned by Major Croghan. A husband of one of the Croghan daughters would stand to inherit some prize property with the best virgin timber from one of Kentucky's wealthiest landowners. Jesup had no property at all and Hancock was a Virginia planter who had met with hard times. The Croghan girl's father owned close to 53,000 acres of land which would have been highly coveted for speculation and worth much more than the same amount of acreage in Virginia was at that time. Major Croghan also owned five houses in Louisville, plus a small town on the Cumberland River. An access to commercial trade on the river, a public ferry boat and other interests must have been very appealing to a Virginia land owner whose land was worn out from raising tobacco. Even the use of the family turnpike wasn't what it used to be since the coming of the steamboat.

Regardless of ulterior motives the suitors might have had when they came courting, Ann and Eliza Croghan were very attractive young ladies. Three young men asked for Ann Jesup's hand in marriage, before and after Jesup appeared on the scene.

The following is evidence that the daughters of Major Croghan were very much sought after:

> George Hancock, Fotheringay, VA to Gen. Jesup, D.C. *Aug. 28, 1820.*
>
> Your having heard that Miss Cn. was to be married to Mr. B., is another instance of the disposition of mankind to circulate reports that have not the smallest grains for their origin. Mr. B. has been only once or twice, I believe, at Locust Grove since the death of Mrs. B. ... I, again, declare that she never would have married D. [Davis].[19]

Geo. Hancock, Fotheringay, *Aug. 6, 1821* to Jesup, D.C

You must have been impressed with the belief that Miss (Ann Croghan) was engaged to D. (Davis). I most positively say that she *never* was engaged to him. At the time you addressed her, I know that she intended to have you, and the morning you addressed her at Mrs. Preston's [Caroline Hancock married Preston], a female friend told her first to consult with her Father and she determined not to encourage you until she had and supposed that she would engage herself to Davis having been acquainted as a college mate with Wm. Croghan and myself, he was introduced by us to our friends. Your friend, George Hancock.[20]

John Croghan, Louisville, *Mar. 11, 1822* to Gen. Jesup Washington City.

Maj Biddle has been here & shortly after his arrival addressed ... She discarded him, nothing satisfied with one repulse, he addressed her a second time; he was of course repulsed & was at the same time informed that she was engaged to you, & would consider it an insult if ever he spoke to her on that subject again. Anderson has been informed that she was engaged to you. He was delighted to hear it for he is very much the friend of both of you. My Father has been very unwell, but is recovering. In haste, Your friend, John Croghan.[21]

Early in the spring, before Ann's wedding, William Croghan, Jr. had made a business trip to Pittsburgh. His time in Pittsburgh extended longer than he had planned, but probably returned before the 17th of May, in time for his sister's wedding.

While in Pittsburgh, a Carlisle University classmate Harmar Denny introduced William Jr. to Mary O'Hara, a daughter of one of the wealthiest families in Pennsylvania and sister of Harmar's wife Elizabeth O'Hara.[22] William fell in love with Mary and was very reluctant to leave Pittsburgh without her acceptance of marriage. Mary accepted his proposal, but marriage would have to be approved by her mother which caused a delay in returning home to Louisville.

William received a "scolding letter" from the Major which reminded him of his neglect of responsibilities in Louisville and said he was to return to Kentucky.

General Thomas Jesup, painting by Samuel King (from the Collection of the Library of Congress)

William Jr. answered his father with the following:

> Could I render you sensibly My DR. Dr. Father! Of the peculiarily of my feelings at this moment in addressing you this letter, you would feel for me, and however great might be your excitement against me for thus neglecting my business, & the services which you would find for me, I am sure you would in charity—forgive me. A recurrence (sic) to the past will sanction this declaration—that although by the laws of our country, I have for several years been, 'my own man' entitled to the privilege of volition, & action—yet—I have never forgotten that I had a father whose advice & consent I have studiously made a condition previous to my entering on any adventure. I have on this occasion this most important occasion, deviated. I have not your consent, your advice will not forgive me for rashness. So sure was I of it I did not seek it. But here I am far from my friends & on a sea of adventure.
>
> My case is simply this: On my arrival in March at Pittsburgh I became acquainted with Miss O'Hara, a daughter of the late Genl. O'Hara of that place. My acquaintance resulted in a warm attached to her & I immediately conceived the idea of addressing her, and no objection was urged, *feeling* and *honor* alike required I should make good my word. With this understanding, she set out for Philadelphia. What could I do?" A [illegible] to feeling were I under these circumstances to have returned home, I would been, truly unhappy, disqualified for business, & had to have returned again to Pittsburgh. Thus it was I determined to meet her at Phila. Many circumstances conspired to delay her arrival. I was here a month before her. Since we have met, we have had many [illegible] some discouraging. I may now congratulate myself, all that is wanting is her Mother's consent, which is written for some days since & may be expected in [illegible] days. On learning the result, I will loose no time in returning home. I hope & Trust My Dear Father this candid [illegible] of my situation, will, (if you have been disposed to be displeased with me,) restore me to your kind & parental consideration. I left with Mr. G. C. Gwathmey $300, Ken money which should you stand in need of. I hope you will make application for. I left a few hundred in the U.S. Bank, also subject entirely to your order. My lave to all the family.
>
> I am sincerely & affectionately Yr. son—
>
> William Croghan Jr.[23]

With Mary's promise of marriage, William Jr. returned to Kentucky, but his wedding was delayed until January of 1823, due to unexpected events that took place at Locust Grove.

When Ann Croghan and Quartermaster General Jesup married on May 17, 1822, the daughter of William and Lucy Clark Croghan would have had no less a grand wedding than the President's daughter. Flowers, ribbons and girls in white dresses threw flowers in the path of the bride and groom. A letter from Lucy's nephew John O'Fallon to Dennis Fitzhugh, dated May 10, 1822 seems to confirm this was a very grand event.

> I suppose on the occasion of so great a match for Ann Croghan there will be a splendid wedding accompanied with much hustle and display. I still think Major Biddle was a superior.[24]

Lucy was a very fashionable lady who ordered some apparel from Paris, as did Dolly Madison and the Jefferson girls, who always obtained their latest fashions from Paris. When new fashions were shown in Paris, France, it would only be a short time until the same fashions would be illustrated in magazines in America. Dolly Madison used an agent in Philadelphia, called a "specialty merchant." In 1811, Major Croghan's former agents, the Gratz brothers, had died. Most likely, he had another agent in Philadelphia who purchased European merchandise for him and shipped by way of New Orleans. One example of this is the ballroom wallpaper for Locust Grove which came from Paris, France, and was designed by Cietti for Reveillon.[25] Only the wealthiest families could afford this wallpaper.

It is not known whether Lucy "rouged" or "snuffed," but Dolly Madison did. "You are aware that she snuffs; but in her hands the snuff-box seems only a gracious implement with which to charm," wrote a female admirer. "Mrs. Madison is said to rouge," said a Ms. Seaton, "but is not evident to my eyes, and I do not think it true." When Dolly entertained, she was never without her lava snuff-box, "a gracious implement with which to charm and many a young man was put at ease when Dolly invited him to help himself."[26]

Strangely enough, in 1833, John Croghan sent his six-year-old niece, Mary (daughter of William Croghan, Jr.) a snuff box.[27] When royalty in England started using snuff before the 1800s, snuff became legitimized and respectable. Most snuff boxes were shaped as a lady's shoe, which was very popular. A tiny silver spoon was always used to sniff the snuff.

When Abigail Adams, who believed in simplicity of dress, was in the White House, she would not recognize a certain lady at one of her soirees, because she rouged and painted her face and she looked like a French lady.

The tradition of the bride wearing white began about 1800, and Lily of the Valley was the wedding preferred flower. The empire style, in fashions had lasted for almost twenty years, but by 1820 the new waistline still existed, following the body's natural contours. Excessive ornamentation of ribbons and flowers were always added to the bride's dress.[28]

The bridesmaids and groomsmen had an amusing time hanging garlands of flowers and ribbons around the rooms. After the supper following the ceremony, the bridesmaids "passed the cake through the ring" and cut slices for the wedding guests to take home, an important tradition of the day.[29]

Bridal couples didn't leave on a honeymoon trip in those days, but stayed at the bride's home. The wedding party also stayed and for three or four days teased the newlyweds unmercifully, a custom that remained in America into the 20th Century.

CHAPTER TWELVE

Eventful Year for Lucy

1822 was an eventful year in Lucy's life. Serena gave birth to St. George Croghan at Locust Grove, on April 23rd , one month before Ann's big wedding celebration in May. That meant space had to be made for George, Serena and their three small children during all the flurry and activities of preparing for the big wedding of the year. Where were all the members of the wedding party and guests housed? Above all, how much rest time was left for Lucy?

This was that dreadful year that malaria swept the Kentucky frontier. Stagnant ponds and mosquitoes made it so unhealthy in Louisville that the city was called, "The Graveyard of the West."[1] Steam boats even refused to stop in Louisville.

Two months later after Ann's wedding, her seventeen-year-old brother Edmund wrote the following to his mother, July 17, 1822:

Dear Mother,

In a day or two we will leave this for the springs. Should I find that my health will permit, I will commence the study of law next October. We had a Commencement. Had I enjoyed any thing like health I would have taken my degree with the last graduates. I will go to the springs, though I am infident (sic) that I shall never recover. If the next October even as well as I now am; let consequences be what they may I will then resume my studies—I shall not return with my brothers, but remain here. Rember (sic) me to all and [illegible] that I remain your affectionate son. E. Croghan.[2]

Edmund, Lucy's youngest child, died on 27th August, 1822. Upon the death of his son, Major Croghan found it necessary to draw up a new will and Edmund's name was removed. Three weeks after Edmund's death, the headline in the Louisville Public Advertiser read:

Major William Croghan, Sr. died 21st September 1822.[3]

Twelve days later, Lucy's sister Ann Clark Gwathmey died at Locust Grove on October 3, 1822.

The lamps must have burned long into the night during that time of sorrow. If, after Ann's wedding, she and Eliza had left Locust Grove with their husbands, they returned very soon. It is possible their brother Edmund's illness and death prevented them from starting on their journey. Both daughters and William Jr. were with their mother in the fall after the death of their brother and father. Lucy's nephew Richard C. Anderson, Jr. described that fateful year best when on his way to serve as the newly-appointed minister to Columbia in South America.

> This year has been a time of affliction to me & to all mine. The deaths of the last year among those who was dearest to me could not occur without striking my heart in a way never to be forgotten. My dear son Louis died on the last day of July 1822. He was the most loved child & I never think of him but with misery. He is buried at his grandfather's near his two sisters & a brother who have gone before him. In October [1822] Mrs. Owen Gwathmey, mother of my Betsy, died after a distressing illness of many months. A better mother & a better woman never lived.
>
> In July [1822] Denis Fitzhugh died; in September Major Wm. Croghan—also in Sept. my brother in law Hezekiah Jones. Mr. Saml. Gwathmey lost two children—one boy of 12 yrs. Old. Isaac G[wathmey] lost his only child. Mr. Tho[mas] Bullitt one—Mr. G[eorge] Woolfolk one.
>
> These were deaths of our nearest relaion. It was the most afflicting season that ever came over the country. The thought of leaving friends, many of whom probably I shall never see again excites painful feelings. My father is old.[4]

This stressful year lingered on for the Croghans. A letter, dated January 1823, from George Hancock in Virginia, inquired of General Jesup's plans to start a trip to Kentucky from Washington, D.C. and mentioned that it had been some time since he had heard from Locust Grove.[5] Hancock was a Virginia State Representative at that time, but did not plan to run for the next session.

On January 28, 1823, the marriage of William Croghan and Mary Carson O'Hara was performed by the Reverend Francis Herron in The First Presbyterian Church of Pittsburgh.[6] The couple lived in Pittsburgh in a townhouse until after their son, William Croghan III was born March of 1824.

The merriment of the matchmaking years were over and the Major was no longer with Lucy. Her new role would now be focused on welcoming new grandchildren into her life.

Events which took place at Locust Grove in 1822:

> Apr. 23—Col. George and Serena Croghan's son, St. George is born.
> May 1—Ann Croghan married Thomas Jesup
> Aug. 27—Edmund Croghan died.
> Sep. 21—William Croghan, Sr. died
> Oct. 3—Ann Clark Gwathmey died at Locust Grove.

That year must have been physically draining for Lucy and quite traumatic, but fortunately Ann and Eliza were still with her. Lucy needed her daughters during her grief and while there Ann's daughter, Lucy Ann Jesup was born in 1823, which must have brightened her life.

The following year, both of Lucy's daughters were at Eliza's home, Fotheringay in Virginia, where Ann's second child, Eliza Hancock Jesup was born in July 1824.

When the New Year 1825 began, America had elected a new President, John Quincy Adams. The "Era of Good Feeling" was over. Andrew Jackson had been a United States Senator from Tennessee, but resigned and left Washington City after President Adams was elected. He was a very angry and disappointed man leaving Washington City after serving two years as U.S. Senator. He went to his grave convinced that Henry Clay of Kentucky had deprived him of the Presidency of the United States by devious means. This wasn't true, but a long-time friendship had ended. President Adams had appointed Henry Clay as his Secretary of State.

Returning home to Tennessee in April 1825, Andrew Jackson, his wife Rachel, and their family were over-night guests at Locust Grove and dined with Lucy, John and other family members and friends. While at Locust Grove, Jackson spoke often of Ann and Eliza who had made a great impression with Washington City society.[7] Evidently the Croghan sisters were present for all important social events after their arrival, especially since the two girls were related to three famous national heroes. In addition, Eliza Croghan Hancock's husband was related to Dolly Madison and Dolly Madison was the "Grande Dame" of Washington society.

The year of 1825 proved to be a year of pomp and circumstance for some members of the family. In May, Lucy Croghan and Sarah Anderson (Colonel Richard Anderson's second wife) assisted in serving at the banquet given for General Lafayette in Louisville.[8] Lafayette's visit to Louisville left the citizens with memories that lingered on for many years.

> In 1825, the Louisville Trustees appropriated $200 to help provide entertainment for General Lafayette, who was arriving on the steamboat "Mechanic." Some miles up the river, the boat struck a snag and went to the bottom in a few minutes. The General sacrificed all his possessions. A boat picked the passengers up and brought them to Louisville for the festivities.
>
> The streets were decorated with flags and streamers and young girls in white muslin dresses threw flowers into the carriage of the smiling Frenchman. The beaux of the town made an elegant picture in their ruffled shirts, broadcloth coats and white linene trousers. The citizens shouted, "Vive La France!" and the General, taking a pinch of snuff, bowed graciously to the crowds.
>
> He was grateful, very grateful that a handsome satin small-clothes, his flowered damask waistcoat and his white velvet coat had not been lost in the wreck. With a small white hand he adjusted his wig to a becoming angle and blew a kiss with Gallic chivalry toward a beautiful young lady, who was smiling at him and curtseying prettily. It was a gala occasion, indeed. This ball for Lafayette, the man who was unquestionably a hero would be the finest party that the town had ever known. French citizens constituted a guard of honor, others assisting were English; Bullitts, Floyds, Taylors, Popes, Prestons, Breckenridges, Thrustons and the Churchills.[9]

William Jr. was back at Locust Grove for a short stay and probably took part in the celebration, but Locust Grove now belonged to him and responsibility for overseeing the farming was left to him.

In the meantime, Ann had received her piano and paintings, sent by Lucy and John, for the new home in Washington, D.C. Nicholas now owned property on the Green River and could have been checking this out before his return to Louisville. Charles was at his place on the river and only John, Lucy and William Jr., were at Locust Grove.

The place on River Road built by Charles was a three-story Georgian style brick house and was a replica of Locust Grove. Today, this stately old home is a Louisville landmark, with an unobstructed view of the river. Blankenbaker Lane on the left of Charles Croghan's house, leads to Locust Grove.

At that time, John was busy with his botanical gardens, medical practice, and "my books to keep me happy."[10]

Lucy was the only Clark sibling living in Louisville. To her, it probably seemed "only yesterday" when her parents and all her family had been together on the banks of the Ohio. Recently, William had arrived in Louisville to escort Frances to live with son John in St. Louis. After a few weeks, Frances died on the 19th of June, 1825 and was buried in beautiful Bellefontaine Cemetery St. Louis, with some of her children.[11] Lucy and William were the last survivors of John and Ann Clark's ten children. Nephews and nieces' lives were also touched.

Richard C. Anderson's diary, "*January 19, 1825*, Since I made the last note, my sufferings have been great indeed. My Betsy, my beloved, my virtuous, my amiable wife left me for Heaven on the 9th of this month about 6 oc in the morning. Her sufferings had been beyond any thing I ever saw. ... Oh God have mercy on me & my children."[12] Betsy had given birth to a son. Both were left in a grave in Columbia.

On August 19, 1825, Lucy's son Nicholas died, leaving his twin brother Charles devastated. William wrote, "On the death of poor Nicholas [Croghan] I felt more than a brothers solicitude for Charles. We were requested to make a journey on his account & forthwith we [William, Charles and Lucy] were off to Washington." These two brothers had been inseparable for twenty-four years. In three short years, many changes had taken place in Lucy's life.

A chapter in Lucy's life had closed when her husband, to whom she had been married for thirty-two years, died. Her years ahead took on a new identity, which brought a certain amount of freedom from so much responsibility at Locust Grove. Lucy would be living with her children, which meant a return journey back up the Ohio River. Forty-one years had passed since Lucy, a young nineteen-year old girl, came down the

Ohio in a flatboat with her family and then her whole life lay ahead of her. It took five months to reach Kentucky when she was nineteen. Returning upstream on a steamboat with her daughters, the steam boat traveled much faster.

When Ann Croghan first arrived in Washington City as the wife of Quartermaster General Thomas Jesup, she felt comfortable with old family connections and acquaintances such as Henry Clay. Kinfolk from Virginia and Kentucky would have welcomed the relatives of the famous Clarks and her brother, Colonel Croghan, was well known among the powerful in the nation's capital. When Lucy and Charles arrived in Washington City, they joined relatives and friends for a historic year of celebrating.[13] The Jesups were known for their hospitality, and some of the prominent leaders of the nation were guests in their home.

Nephew Richard Anderson wrote about seeing "Aunt Croghan" in Washington, "*Sept. 8, 1825,* There I saw at Genl. Jessup's Aunt [Lucy Clark] Croghan & Charles C[roghan]. To see that old woman in Washington declared the changes & reverses which take place in this world."[14]

Lucy was present for social events in the president's house. President Monroe would have made her welcome. After all, he had been guest in her home. Lucy's son-in-law Thomas Jesup was like a son to President Monroe. Lucy was recognized as the sister of two heroes and the mother of one. Dolly Madison made her welcome to Washington Society. Henry Clay would have taken Lucy's arm and proudly introduced her to many important people.

Four years later, another Presidential inauguration took place on February 28, 1829. Governor Clark wrote from Washington to son Lewis at West Point, NY.

> Mrs, Gen. Floyd also with her daughter, William & George [Croghan] and hundreds of strangers and others coming in to see Gen. Jackson's enaugration.[sic] The girls are admired and they as well as your Mama [stepmother] spent their times pleasantly.[15]

Lucy, as well as many of the Clarks from St. Louis, was also in Washington in the winter of 1832–1833 for Jackson's second inaugural. Serena Livingston Croghan's Uncle Edward Livingston was part of Jackson's cabinet and the Livingston aunt spent many hours with the Jacksons in the White House family quarters.[16]

When Lucy returned after her first visit to Washington D.C., she was back at Locust Grove with William, Jr. and family in August 1827. George was causing worries for the family. Before he had decided to live in New Orleans, the Livingstons had tried to persuade George to settle down with a job near them and probably were very disappointed when he accepted the position of postmaster in New Orleans. John urged George not to accept the job as postmaster in the "vile" corrupt city.[17]

George must have found it difficult to resist the lifestyle which his wealthy friends and the Bullitts, Clays, and Livingston relatives were living in New Orleans. Serena would not live there and before long George realized he could not support himself and his family on his income. He was appointed Inspector General of the United States army in December 1825, but not before he was in deep financial trouble.

Serena's brother, John Livingston, traveled to New Orleans to represent George, who needed legal assistance after becoming indebted to the post office and on the verge of losing his plantation. Colonel Croghan and John Livingston had fought together with General Jackson in New Orleans. On his way to New Orleans, Serena's brother spent the weekend at Locust Grove.[18] In August 1827, William Croghan wrote to Jesup. "I never knew Mama to be so low spirited as she then was on George's account, she insists on it she will never see him again."[19] About this time, George began to show signs of a drinking problem.

New Orleans had been the scene of another sad loss for Lucy when on November 8, 1824, her nephew, John Gwathmey, died there about the same time that Colonel George Croghan arrived. Richard C. Anderson, Jr. wrote in his diary, Feb 19, 1825, "Poor man—he had an unhappy life. The last years were most miserable but should not have been so, as he preserved his honesty & character."[20]

Lucy's nephew, R.C. Anderson Jr., had returned with his children to Kentucky after Betsy died (January 1825) in Colombia, South America. He had found a school for his daughter but, another concern was his father's debts. He soon lost Soldiers Retreat to creditors. R.C. Anderson had kept his father afloat, but "I find nothing has been done about my father's business. My responsibilities still continue for him & I see no way of clearing myself soon."[21]

Betsy's sister, Diana Gwathmey (Thomas) Bullitt, met him at the dock when he returned. Diana came alone, since her husband had recently died. Her son Ferdinand, who had gone to Colombia with Anderson as his private secretary, had also died in Colombia.[22] The trip to Colombia was supposed to be for financial gain to cover indebtedness. In the end, this scheme ended in great losses for other family members in more ways than one. At this time, the third Clark generation had begun to show signs of how true the old adage was, "easy come, easy go."

CHAPTER THIRTEEN

Last Days at Locust Grove

In October 1827, William Croghan's, beloved Mary died and was buried at Locust Grove. Two weeks later, Mary's sister came from Pittsburgh to accompany William and his two young children back to live with the O'Haras. William returned to Locust Grove, but six months later, was back in Pittsburgh after his five-year-old son died in April 1828.[1] He was at a loss in knowing what to do, but decided to stay in Pittsburgh and start managing his wife's Pittsburgh property and investments, which she inherited from the O'Hara fortune. It was quite a struggle at first, but with his inherited business sense, he ultimately turned the holdings into millions.

William sold Locust Grove to George and Eliza Hancock. The next three years there was much visiting between Lucy and her children, who lived in Washington, Pittsburgh, New York, and Locust Grove. But the summer of 1832, Lucy was living with Ann in Washington.

John and Charles sailed for Europe, hoping the change might help to regain his health. Attending Charles was very confining for John, since this was his first trip abroad, and he wrote of his desire to stop off at interesting places. As the ship passed the shores of Ireland, he longed to go ashore to the land of his father, who had talked so many times about his homeland.

London, Aug. 18th, 1832

My Dear Ann,

Our passage across the Atlantic was unusually pleasant & quick The first land that we saw was the 'Irish Coast.' You cannot imagine the sensations I experienced in beholding the native land of my Father: a land so intimately interwoven with all my early associations. ... you have a view for some distance of the coast of Ireland, and on the right of the Welsh mountains with cottages innumerable. John Croghan.[2]

114

Dr. John Croghan (From the collection of Historic Locust Grove, Inc.)

In October, John informed his family that Charles died on October 21, 1832. He was buried in Paris.

> He died on the night of the 21st and was interred in the cemetery of Pere La Chaise on the 23rd. Attentions both medical and otherwise were shown him while living and his corpse was followed to the grave by nearly all of the Americans in this city.[3]

Lafayette, James Fenimore Cooper and other famous Americans living in Paris were part of the funeral party to the cemetery. John decided to stay in Europe and traveled for several months.

In the meantime, George and Eliza Hancock were living at Locust Grove, where the ponds and swamps needed draining to eliminate cholera that was always a threat. In June 24, 1833, the Jesups and Lucy in Washington received a letter from George Hancock,

> Eliza is better today. Mrs. Pearce is with her, [Jonathan Clark's daughter] the Doctr. thinks her convalescent. The cholera is with us; we have 5 cases today—it yields readily to medicine and I hope none will prove fatal. You can conceive nothing to equal the gloom spread over the country here. No one leaves home. Crops of wheat standing uncut, corn fields abandoned to winds. and what makes all worse is it is incessantly raining; as yet there are few cases in Louisville.[4]

> Locust grove June 30, 1833
> Dear Genl.

> Since my last to you the cholera has increased to an alarming extent. We have not well ones enough to attend the sick—and it is difficult to get a Physician. Dr. Tompkins is with us now and has promised not to leave us until Eliza is better. I fear her situation is *very critical*. My mother was taken with cholera (I fear) tonight. If Dr. Croghan is with you for Gods sake sent him on to us.[5]

Nine months after Charles died, Lucy lost Eliza, one of her beloved daughters. While in New York, William canceled a trip to Quebec upon learning of his sister Eliza's death and rushed to his mother in Washington. William had just arrived, when John also, returned from Europe. William and Ann stayed near and comforted their mother.

The family didn't return to Kentucky until one year later, July 1834 which was a sad journey. It must have been especially difficult for Lucy when Locust Grove came into view. Two of her children would no longer be there and she had a new grave to visit.

William wrote to the Jesups, "On yesterday morning we reached here, finding all concern well & impatient to see Mama. Her health improved every day since she commenced her journey. Already she appears at home here."[6] After one month at Locust Grove, she returned to Washington, but George Hancock sold Locust Grove later in 1834. It was John who purchased the old home which had been in the family for almost fifty years. Lucy came back to Locust Grove and she never again traveled further than Louisville, so her remaining three children and their families often visited Locust Grove.

When William Croghan sold Locust Grove to his sister Eliza and George Hancock August 1831, the Hancocks had insurance coverage for the three-story house dwelling for $4,000; the private library for $1,500; and household furnishings for $1,500.[7] When John purchased the house in 1834, he and Lucy returned to the old home place.

On March 4, 1835, John found it difficult to write to Ann. He had complied with his mother's request to come sit in her room while writing. After a while, John complained in the letter that writing was very difficult because "Mother has talked, incessantly while she sews on a rug." On Lucy's last visit with Ann in Washington, Lucy's slave, Alfred, had disappeared. Ann's husband, General Thomas Jesup, was sometimes preoccupied with finding runaway slaves and then contacting their former master. He found Alfred and wanted to know what Lucy wished to do with him. She took the pen and wrote her own message at the end of letter. It is very clear and to the point.

> P.S. Dear General, by your letter to John, I find Alfred is unwilling to return to Ken: I, therefore, have no objection as he evinces so little gratitude, and as his conduct has been so improper to dispose of him. Affectionately Yours, Lucy Croghan.[8]

Many southern women found slavery a burden and felt a certain amount of freedom when the slave was free to live their own life. Alfred did return later and letters show he spent his last days at Mammoth Cave.

117

While General Jesup was fighting to capture the Indians in the Everglades, Ann and children had returned to Locust grove and living with her mother and John. They were there for almost four years. For Lucy, perhaps the stairs were becoming more difficult to climb and a bedroom was built for Lucy adjacent to the family kitchen. John wrote, "Mother, Ann and children having gone to the city, I am building a small room."[9]

There had been no bedrooms on the first floor until the library was converted into a bedroom for the nine years General Clark lived there. After his death, the General's room had been returned to its former use, a library with many books and fine pieces of art. William Croghan, Jr. had inherited Locust Grove and all these furnishings.

The room which had been built for Lucy was removed during the 1960's restoration. Lucy was expunged from the history of Locust Grove as not having an important role, even though this was Lucy's home from 1790 to 1838.

Lucy died on April 4, 1838, but not before she made provisions in her will for her daughter-in-law, Serena and her three grandchildren. Although, some had turned against Serena, Lucy realized what a difficult time was in store for George's family.

John and Ann Clark's last child died in 1838. Lucy died in April and William died on September 1, 1838. The Clarks were a remarkable family that would be remembered by historians for years to come.

Major E. A. Hitchcock, Deputy Agent of Indian Department, St. Louis, made the announcement for the nation:

> The venerable General William Clark died between the hours of 7 and 8 last evening at the house of his son, Lewis, in this city.
>
> The funeral was a large military one and very imposing; a company of soldiers escorted the remains from the residence of Meriwether Lewis Clark, where the General had died, out to the estate of Colonel O'Fallon on the Bellefontaine Road, where they were placed in the family vault until after the opening of Bellefontaine Cemetary many years later. People lined the street for blocks to watch the cortege led by a military band. Next came the S. Louis Greys in full uniform, followed by the Masonic fraternity, and the hearse drown by four white horses in black plumes and trappings, each led by a 'Negroe'. Last, before the carriages which held the family and friends, the General's horse was

led by his servant, both in black. Clark's pistol, bolster, and sword were laid on the saddle and his spurred boots were reversed in the stirrups. Following the carriages were many men on horseback; and, as we came within a half-mile or so of the burying ground, minute guns were fired from a cannon. So this great and good man, whose whole life had been given in selfless service to his country, was laid to rest.[10]

Ann Jesup and her children were living at Locust Grove with John when Lucy died and were there when they heard the news of Uncle William Clark's death.

Many years later, a monument was placed on William's grave.

On a hillside overlooking the Mississippi River a handsome monument, placed there during the celebration of the Louisiana Purchase Exposition at St. Louis in 1904, now covers the grave of the great explorer and his second wife, Harriet Kennerly Radford Clark, as had been directed by their son, Jefferson K. Clark, before his death in 1900.

The unveiling of this monument in the Clark plot of the cemetery was an inspiring ceremony witnessed by five generations of General Clark's descendants and a gathering of distinguished people from all parts of the country and even from England [Mary Schenley's daughter, Lady Ellenburgh was visiting in St. Louis, at that time].

Mounted against the tall shaft of Missouri granite, Clark's bronze bust (by William Ordway Partridge) faces the Mississippi River, and his sculptured eyes look across toward the point from which he and Lewis started on their epoch-making journey through thousands of miles of savage wilderness to plant the flag of the United States on the shores of the Pacific. On the face of the monument is inscribed this fitting Biblical quotation: 'Behold, the Lord thy God hath set the land before thee: Go up and possess it.'

During the ceremonies, as 'The Rosary' was being softly played by the band from Jefferson Barracks, I realized that among those there assembled in memory of this great man, I was the only one who had ever known him or clasped his hand.[11]

*William Clark Monument in Bellefontaine Cemetery, St. Louis, photograph by Emil
Boehl, ca. 1902 (courtesy of Missouri Historical Society)*

At the time of Uncle William Clark's death, William Croghan, Jr. was liv-
ing in Pittsburgh and Colonel George Croghan was at Fort Snelling in
Minnesota on an army inspection tour. The four oldest Croghan children,
John, George, William and Ann, were the only Locust Grove Croghans left.

John Croghan's remaining years were spent in pursuing an enterprise
that would receive worldwide attention and long-lasting interest, when he
established and operated Mammoth Cave as a worldwide tourist attraction. During that time, John allowed Locust Grove to become run down
because he had sent almost all the slaves to work at Mammoth Cave. Even
"ungrateful Alfred" had been sent to the cave and must have died there.
Some of these Locust Grove slaves lived and died in the area of Mammoth
Cave and today it is very possible their descendants live in the area. There
were Croghans who lived in Western Kentucky and there are those who use
the spelling, "Cron."

Some may say Colonel George was a failure after his first brush with fame in 1813. The following years while traveling to military forts on the frontier, he had begun to drink, but was still committed to improve life for the soldiers who were far removed from society. His official Army reports of inspection on the Western Frontier can be found in the National Archives. These reports were sent to Washington, D.C., where they served as a valuable tool in building military policy and made the careers of two Army officers, General Thomas Jesup and General Scott, who read and acted upon his recommendations. These official reports place Inspector Colonel George in a different light for the reader.

William Croghan Jr. became a millionaire. Ann Croghan Jesup died in 1846, leaving six children whose ages ranged from six to twenty-seven.

Generations of granddaughters would follow in Lucy's footsteps. These women would be tested as they overcame obstacles in their lives, but they were indomitable. They were well educated in the arts, literature and very active in community causes. They lived far and wide from their pioneer grandmother's Locust Grove, but her influence extended beyond that grave in Kentucky. They always remembered their Croghan ancestors at Locust Grove and were proud of their Southern heritage.

For many years there were no Croghans at Locust Grove. In Dr. Christopher Graham's letter to Lyman Draper in 1882, he spoke of "recently been over the Croghan farm hunting fossils. No one now living there has any knowledge of the former occupants."[12]

The next year, in July 1883, Colonel Reuben Durrett made a similar excursion. From the Taylor monument some excursionists went across the country to the banks of the Ohio to visit Locust Grove.

> Here stands a grand old family mansion erected by William Croghan, who came to Louisville in 1784 and died in 1822. Wm. Croghan married the sister of Gen. Geo. Roger Clark, and it was here that Gen. Clark lived after the accident by which he lost his leg, 1809.

> Locust Grove is no longer the property of the Croghans. A few years ago purchased by the widow of Capt. J.M. Paul, who now occupies it. It bears but little resemblance to the abode of the Croghans three-quarters of a century ago. The old mansion is there, with it's walls of brick, without a fracture, and it's woodwork of

121

walnut lending a solid appearance to rooms large enough for modern parlors, but all else is changed. The locusts that gave name to the place have felt the force of storms and the decay of years. The lawns have put on the garb of fields of culture, and the old family graveyard where sleep our pioneers is a tangled thicket through which none may creep to read the inscriptions which tell of the honored dead.[13]

This honored place called Locust Grove would be remembered by nieces, nephews and neighbors as the place of happy times. One nephew, multimillionaire John O'Fallon in St. Louis, learned of the deaths of John and George and wrote to Zachary Taylor's brother, Joseph, that he had always loved Locust Grove and wished to own the old place for one of his sons. John O'Fallon's dream was to be buried there with the Croghans, but his wife would never move to Kentucky.[14] Joseph answered:

> Well I should like to see you once more and talk over our boyish days as there are but few of us left now, of the many boys who were our playmates in our young days for death has thinned them off to a fearful extent—it is melancholy to think of the large family left by your Uncle Major Croghan & but one is left of them.
>
> J P Taylor.
>
> PS. I went to Washington City to witness the Inauguration but such was the wish of office seekers, that I was so disgusted that I left the next day.[15]

One month later, Joseph's brother President Zachary Taylor was dead.

Ten years later, after Croghans were no longer at Locust Grove, grandchildren and great-grandchildren of John and Ann Clark were torn apart during the worst war in which the United States ever participated. Life was cut short for young cousins. Some fought for the Confederacy and others for the Union and some never spoke to each other again. It truly can be said that the era of 1765 to 1865 was "the . . . best of times . . . the worst of times" (Dickens).

The Revolutionary War gave birth to a new nation and the Civil War held it together.

PART II

Born to Be a Soldier

CHAPTER ONE

The Livingstons

When Colonel George Croghan brought his new bride, Serena Livingston, to Locust Grove, June 1817, it was a blending of Hudson River Dynasty with Virginia Tide Water Gentry in Kentucky on the Ohio River. All relatives and friends had arrived at Locust Grove to meet the niece of Chancellor Robert Livingston, the highest judicial officer in New York, a Delegate to the Continental Congress and a member of the drafting committee for the Declaration of Independence. He was signer of Louisiana Purchase, gave oath of office to President George Washington and was a member of the New York convention that ratified the Constitution.[1]

Philip Livingston (Serena's Grandfather's first cousin) served in the Continental Congress (1775–1778) and signed the Declaration of Independence.[2] Her Uncle Edward Livingston was Mayor of New York, Secretary of State, and Minister to France. Aunt Janet Livingston married Revolutionary War hero, Brigadier General Richard Montgomery.

Aunt Alida Livingston married General John Armstrong, who was in command on the Potomac River, War of 1812, when the British burned the White House. While Serena was growing up, she must have heard many personal stories and gossip concerning these famous people.

This New York "Belle of the Hudson" came from a world of magnificent manor houses. The Livingstons became a landed aristocracy as a result of acquiring and adding property along the Hudson River when Serena's Grandfather, Judge Robert Livingston of Clermont (1688–1775) married the only heir to this property, Margaret Beekman, which added nearly 200,000 acres to his 160,000 acres along the Hudson from Rhinebeck to Tivoli. Judge Robert and Margaret Beekman Livingston had ten children who built manor houses on a fourteen mile stretch which gave a sweeping view along the Hudson even into Dutchess County.[3]

Serena's father John R. Livingston built more than one house on the Manor, but the showplace was *Messena* which became famous for its lofty, glass-domed library ceiling. One of his houses, *Edgewater*, still stands today with its sweeping view of the Hudson. Serena's youngest sister, Margaret, who married Loundes Brown, inherited this estate. Livingstons owned land in the East Indies. The international trade in rum and sugar, a triangular trade involving Africa, the West Indies, and the colonies was started by Margaret Beekman Livingston's father, who had acquired his wealth in slave trade.[4]

Serena's father John R. lacked the drive for securing a place in history as a war hero or seeking a high position in government as had other males in the family, but a more shrewd business man than John R. Livingston was hard to find. His desire for wealth sometimes superceded his scruples but unlikely he ever crossed into any illegal practices.

Even though John R. Livingston was anxious to be a rich man, he took his place beside his brothers and other family members in defense of his country. In 1777, July 14, he went off to Fort Edward as a volunteer.[5] Before leaving, John R. "was engaged in a flutter of rum with his partners, Walter and John Carter of Albany, New York." He instructed them "if any accident happens, you must have an eye to the rum at Clermont," Livingston manor house.[6] John Livingston soon disposed of his rum interests, and his mind was very soon turned towards new business ventures. This pattern of seeking his fortune was always present which proved to have its rewards. A former business partner of John Robert Livingston's had this to say about him, "to John R. Liv., War was simply a means of making money."[7] In 1780, John R. wanted to go set up a residence in Holland to drive a trade between Holland and Saint Eustatious, "a plan that cannot fail to make a fortune." Even Robert Morris, who was "morally unsound" told John R. that this would involve trading with the enemy. When John still insisted on going to London, his brother, The Chancellor, firmly told him to give it up. John R. replied, "Poverty is a curse I can't bear, with it a man had better not exist and you must know that the family are too much distressed to give me any assistance."[8]

The Chancellor's rebuke to his younger brother did not curtail his self-interest and John never failed to express his aim to make money. He suggested to the Chancellor to go for the presidency of Congress, "it will be," he said, "productive of great advantages to your friends."[9] In July, 1778, John R. Livingston was into "something to do with the purchase of Loyalist goods in New York, in the advance of the capture of that city."[10]

Judge Robert and Margaret Beekman Livingston

(Serena's Paternal Grandparents)

A. **JANET,** born August 27, 1743, married.

Brigadier General Richard Montgomery, July 24, 1773; Died November 6, 1828.

B. **CATHARINE,** born February 20, 1745; Died April 29, 1752.

C. **ROBERT R.** born November 27, 1746 was oldest son; married Mary Stevens, September 9, 1770; was Chancellor of New York in 1777; appointed Secretary of Foreign Affairs in 1781; administered oath to George Washington on 30th April 1789; Minister to France 1801 and with James Monroe, signed the Louisiana Purchase; Died February 26, 1813.

D. **MARGARET,** born January 5, 1749; married February 22, 1779 to Thomas T. Tillotson, Physician and Surgeon-General, Northern Department; Died March 19, 1823.

E. **HENRY BEEKMAN,** born November 9, 1750, second son, married March 11, 1781, Anne Hume Shippen; Died November 5, 1831.

F. **CATHARINE,** born October 14, 1752; married June 30, 1793 to the Reverend Freeborn Garretson; died July 14, 1849.

G. **JOHN R.** (Serena's father) born February 13, 1755; (1) married Margaret Sheaffe, 1779 Margaret died 1784 and there were no children; (2) married ELIZA McEvers, (Serena's mother) May 30, 1789. John R. Livingston died September 1851.

1. **ROBERT MONTGOMERY LIVINGSTON,** born 1790, New York, m. Sarah Bache of New York and was living in NY in 1828. He died Jan. 1838 in N.Y. They had three sons and three daughters:

2. **SERENA LIVINGSTON** born 25th August 1794 married May 1817— Col. George Croghan of the U.S. Army. He was born 15th Nov. 1791 in Kentucky, and was Captain, Major, Lieutenant Colonel and Colonel in the U.S. Army. He received from the Congress of the U.S. a gold medal, for his gallant defense of Fort Stevenson, Ohio in the war of 1812. In 1825 he was appointed an Inspector General in the U.S. Army, and he served under Gen. Taylor in the War with Mexico. He died 8th of Jan. 1849 at New Orleans.

3. **ANGELICA LIVINGSTON** born 1795 or 1796 New York, died 1815 unmarried in France, age 19.

4. **EDWARD LIVINGSTON** married Sarah Sukely, dau. of Geo. Sukely, merchant of New York.

5. **CHARLES LIVINGSTON,** born at New York, unmarried.

6. **JOHN ROBERT LIVINGSTON, JR.** graduated at Princeton in 1822, and was a lawyer. Feb.27, 1827, at Trinity Ch. Parish, New York, John married his cousin Mary McEvers of New York where he settled and was Corporation attorney and afterwards Navy agent. He died 1871—had 2 sons and 1 daughter.

7. **ELIZA LIVINGSTON** married Benjamin Paige.

8. **MARGARET LIVINGSTON** born at New York, married Captain Lowndes Brown of the U.S. Army, had a daughter, Elizabeth Lowndes. Margaret inherited from her father, *Edgewater.*

H. **GERTRUDE,** born April 16, 1757; married Morgan Lewis, May 11, 1779; died March 9, 1833.

I. **JOANNA,** born September 14, 1759; married Peter R. Livingston (her cousin); Died March 1, 1829.

J. **ALIDA,** born December 24, 1761; married John Armstrong, January 19, 1789, who took a lot of criticism when the White House was burned by the British in War of 1812. Their daughter, Margaret married William B. Astor, son of John Jacob Astor, famous fur trader. Armstrong-Napoleon.

K. **EDWARD,** youngest son, born May 28, 1764 was Mayor of New York and also served as a six-term congressman (while living on his plantations in Louisiana). President Andrew Jackson, an old friend of Edward, appointed him to the second place in the United States, Secretary of State in 1831. Two years later (1833), President Jackson appointed Edward Livingston, Minister of France. Edward married (1st Wife) Mary McEvers (sister to Eliza McEvers, his brother John's wife), April 10, 1788, (2nd) Marie Louisa Valentine D'Avezaac Castra Moreau, June 3, 1805. Edward Livingston died, May 23, 1836.

CHAPTER TWO

British March on Livingston Manor

Colonel George and Serena had grown up listening to the older ones relive those war years and how it had effected their lives. Their children heard stories about both sides of their Grandparents Croghan and Livingston. Although, the Clark family in Virginia had lost two sons, their home was saved from the ravages and destruction of war. Serena's family came face to face with the devastating destruction that war brought when the British marched through and burned their homes.

The Livingston world of peacefulness and gracious living was interrupted by the guns of war in their valley when Burgoyne's Army marched down the Hudson River. Margaret Livingston and daughters had no men folks to defend them on July 6, 1776. They were all away fighting. Her son Chancellor Robert received a letter expressing her fears and warns against attempting to come near home. The "four thousand" Tories or Loyalists, were nearby hiding in the woods.[1]

There were sleepless nights according to some citizens and descendants. Years later, they could remember lying in one of Clermont's rooms and heard all night the rumble of the artillery and the tramp of men and of horses when Washington passed this way to unite with the Frenchmen in an advance upon New York. Found in General Washington's diary, he "halted for rest at the old Dutch Church, which is opposite the manor-house grounds."[2]

After Washington's defeated army left New York, an immediate fear of perilous danger set in. Margaret Livingston and her daughters made plans to escape to Salisbury, but throughout the summer of 1777, silver and porcelain were buried in the woods, books lowered into dry wells covered with rubbish and French mirrors concealed in false walls. Leaving in coaches, looking back and seeing smoke as they went over the hill, Margaret, daughters and servants knew they had just escaped in time.

The British had started burning Clermont (Serena's Grandfather's house) and Robert Livingston's Belvedere.[3]

Had "Mama" Livingston, daughters and household not escaped when they did, most likely they would have met the same fate as other women and children who were not as fortunate as the Livingstons. British orders were to burn all houses. "The British burned the houses, stripping the women and children, and leaving them exposed to the inclemency of the weather; while the men were carried away as prisoners, to languish in the old sugar House in New York, or to die in one of the pestilential prison-ships, in the Wallabout basin."[4]

It was lonely and dreary at Christmas 1777 for Margaret with her daughters and young Edward in Salisbury. Other family members were present at "cousin" Robert Livingston's [not the Chancellor] manor house. Upon learning that, "John R. Livingston was in Boston reaping enormous profits from the war, they immediately planned a trip to Boston to see what opportunities lay there for them."[5]

In the spring, April 11, 1778, Chancellor Robert wrote from the Manor of Livingston to his brother John R. that Mama had left that morning to return to Clermont. She had put up a hut and spent the greater part of the past week there. When he arrived two days later and as he went over the hill and saw the ruins, he tried to describe his sensation and how extreme-ly melancholy it made him to see the ruins and the top of the chimney. Always before, this scene had excited him with the most pleasing emotions and then he remembered the joys of that once social fireside. As he came down the hill and saw his Mama in her garden, "tending with solitary care those plants her hands had reared in happier times ... Good Gods what a flood of tender thoughts burst in upon me." He didn't know how to describe how much he was "unmanned" to speak to her for some time.[6]

Mama Livingston seemed to be a lost soul, she had no pleasure any-where else or conversation. She only had one interest which was rebuild-ing and she wandered around like a ghost. He could hardly gaze upon the loss of his house and all the luxury in ruins and some pleasing tender scene would be recalled from the past

While in Boston, John R. [Serena's father] was courting the young ladies, "they are very handsome (he once wrote) but very awkward and ignorant, so that I am in little danger of getting married among them".[7]

However, romance and love took over when he met Miss Margaret Sheaffe who had no dowry, and her mother (though moving in the first circles) had been reduced to keeping a shop. Margaret Sheaffe's father was Sir Roger Hale Sheaffe, Baronet, a Major General in the British Army. John and Margaret married 1779. She died in 1784 leaving no children.[8]

Ulster, Kingston—on 30th of May 1789, John R. Livingston married Eliza McEvers, sister of his brother Edward's wife. Eliza and sister Mary were the daughters of Charles McEvers a merchant of New York. The following is one verse written for the McEvers sisters by Edward Livingston, who had aspirations of writing great poetry:

> The perfect beauty which you seek,
> In Anna's verse I find:
> It glows on fair Eliza's cheek,
> And swells in Mary's mind.[9]

These three celebrated McEvers sisters, from all accounts, the compliments of their beauty and accomplishments were not empty flattery. Eliza was to become the mother of Serena.[10]

In 1798, a young friend was visiting Serena's Uncle Armstrong and had this to say:

> The trunk of the Livingston family is wide, it spreads its branches in all directions. All of these people are well-to-do, polite, related, closely knit in friendships and sentiment—aristicratic but good-hearted people. He mentioned in passing their custom of calling each other 'not by name but by title, even the women.[11]

This custom held true for many generations later, which helps to explain and understand why Serena Livingston Croghan was referred to as "Mrs. Croghan" by her grandson, George Croghan III (last owner of Locust Grove) in his letters.[12]

Serena's Aunt Alida and Uncle John Armstrong, while in France, were in the company of France's leaders and diplomats. They attended Napoleon's Coronation, as one can see in the famous painting of the coronation which now hangs in the Louvre. John Armstrong is standing in the top row (first left) behind Napoleon. Serena probably heard some interesting gossip about these famous people during the drawing room

Serena's mother, Eliza McEvers Livingston (courtesy of New York State OPRHP)

gatherings, especially at Christmas time when all Livingston aunts, uncles and cousins were together at Mama Livingstons.

Very fascinating discussions would certainly have taken place when Serena was growing up. The visits of General and Martha Washington to Clermont and other "retold stories" would not have gone unnoticed by Serena. On June 4, 1782, Margaret Beekman wrote to her son, Chancellor Robert Livingston.

> Yesterday morning the Genl. and his Lady left Clermont & returned to headquarters after calling upon Harry on his way down we happened to have Mr. & mrs. Provost to dine with us when news was brought that ye Genl. and Gover. were at the Landing—the whole soon after entered to my great joy—the Genl. said he was going with the govr. to Albany and would leave his Lady with me till his return which he did the next day at Ten o'clock—he admired the place and Mrs. Washington seemed pleased we went to the Manor spent a day with peter & the next morning returned home after loging at Walters, red hook Church was crowded to see Mrs. Washn. Dr. Livingston preached she was here eight days.[13]

Serena and her sisters lived in what Washington Irving referred to as "fairyland." Irving was a frequent visitor at the Livingstons and wrote about one of his visits with Serena's family on the Hudson River in the summer of 1812.

> In August, I proceeded to the country-seat of John R. Livingston, where I remained for a week in complete fairy-land. His seat is spacious and elegant, with fine grounds around it, and the neighborhood is very gay and hospitable. I dined twice at the Chancellor's and once at old Mrs. Montgomery's. Our own household was numerous and charming. In addition to the ladies of the family there were Miss McEvers and Miss Hayward. Had you but seen me, happy rogue, up to my ears in "an ocean of peacock's feathers," or rather like a "strawberry smothered in cream"! The mode of living at the Manor is exactly after my own heart. You have every variety of rural amusement within your reach, and are left to yourself to occupy your time as you please. We made several charming excursions, and you may suppose how delightful they

Serena Livingston Croghan, ca. 1815, painting by Thomas Sully (from the collection of Historic Locust Grove, Inc.)

Angelica Livingston, ca. 1815, painting by Thomas Sully (courtesy of New York State OPRHP)

were, through such beautiful scenery, with such fine women to accompany you. They surpassed even our Sunday morning rambles among the groves on the banks of the Hudson.[14]

Serena was eighteen years old that summer and Angelica was sixteen. In 1815, both sisters sat for portrait painting by Thomas Sully. The two sisters have the same pose, same hair style, same dress of that period and the same harp painted in the foreground. Serena's portrait, which now hangs in the state dining room at Locust Grove was a gift from Serena's great grandson Donald Newhall.[15] Angelica died very soon after the painting was finished. She had been taken to France hoping to improve her health, but died in Marseilles sometime before May 1817.

Henry Lee (older half-brother of General Robert E. Lee) was part of the Livingston scene on the Hudson River in those days and family members have written about Serena's love for Henry.[16] Strong ties had always existed between the New York Livingstons and the Lees of Virginia beginning with the struggle for freedom from the British. The 1810 census records show Henry Lee of "Stratford" Westmoreland County, Virginia, was living in Columbia County, New York near the Livingstons and Serena's Uncle John Armstrong, considered Henry Lee to be, "one of his confidents."[17] During the War of 1812, Henry fought in Upstate New York and afterwards he and the Livingstons were with Andrew Jackson in New Orleans.[18]

While Henry Lee was soldiering in New York, there was another battle taking place on the Western front called Fort Stevenson, Ohio. A young twenty-one-year-old boy from Kentucky with one hundred and sixty men and only one gun, a six-pounder, defended the fort against 500 British regulars and more than seven or eight hundred Indians under the great chief Tecumseh. George Croghan of Kentucky became a celebrated hero, and future husband of Serena Livingston.

CHAPTER THREE

Born to Be a Soldier

In George Croghan's youth and early manhood, the future looked bright. His favorite sports were shooting and fox-hunting. At times, he would start at midnight or shortly after into the forest, alone or with his negro boy attendant, to chase a fox or to seek other game. George liked to read, particularly history and biography. A Croghan schoolmate wrote in 1815 that George Croghan never tired of hearing the tales of war and battles. "In school, his selections of speeches were always of a martial cast. A disrespectful remark about George Washington was fighting words. Shakespeare, he admired and could recite most of the famous passages."[1]

He was raised as a young gentleman of Virginia ancestry, and received a Bachelor of Arts degree, from College of William and Mary. The subject of his graduating oration was "Expatriation" for which much interest was shown at the time, but this topic had a special meaning for George Croghan.

George remembered well what happened to his father after the Revolutionary War ended. Major William who fought with The Virginia Line was not allowed to keep his land only because no improvements had been made while fighting in the Revolutionary War. The subject of "Expatriation" was familiar to George and clearly understood.

The schoolmate continued:

> He was remarkable for discretion and steadiness. His opinions, when once formed, were maintained with modest but persevering firmness; and the propriety of his decisions generally justified the spirit in which they were defended. Yet, though rigid to his adherence to principles, and in his estimate of what was right or improper, in cases of minor importance, he was all compliance.

Col. George Croghan, painting by John Wesley Jarvis (from the collection of Historic Locust Grove, Inc.)

I never met with a youth who would so cheerfully sacrifice every personal gratification to the wishes or accommodation of his friends. In sickness or disappointment be evinced a degree of patience and fortitude which could not have been exceeded by any veteran in the school of misfortune or philosophy.

He is (as his countenance indicates) rather of a serious cast of mind; yet no one admires more a pleasant anecdote or an unaffected sally of wit. With his friends, he is affable and free from reserve; his manners are prepossessing; he dislikes ostentation, and was never heard to utter a word in praise of himself.[2]

In 1811, Governor of Indiana Territory, William Henry Harrison called for volunteers to defend the frontier against Chief Tecumseh. Three Clark cousins, John O'Fallon, John R. Gwathmey, George Croghan and other Kentuckians were quick to heed the call for Harrison's army.[3]

Before the decisive battle of Tippecanoe, George Croghan's handsome appearance and intelligent discharge of his duties had attracted the attention of the officers and he had been made an aide-de-camp to General Boyd, the second in command. During the battle the young Kentuckian rode from post to post cheering the men and saying; 'Now, my brave fellows, now is the time to do a main business. A cant phrase among the soldiers on the Tippecanoe Campaign was "to do a main business."[4]

Generals Harrison and Boyd recommended a captaincy for George in the Seventeenth United States infantry regiment, when a prospect of war with Britain appeared in spring of 1812. During August, one garrison was ordered to march to the relief of Fort Wayne, Indiana and then down the Maumee River to Fort Defiance in Ohio. Captain Croghan had won the respect of his superiors while on the march, "by showing his skill with which he selected and protected his camping places."[5]

There, Captain Croghan was left in command at Fort Defiance. The troops on the expedition to Fort Wayne had very little food, were badly equipped for winter and suffered great hardships and then massacre on the Raisin River. Captain Croghan would have been on the expedition had Harrison not left him behind in command of Fort Defiance.

Even though George was twenty-one his excellent record in the defense of Fort Defiance and Fort Meigs led Harrison in making the decision to give him command of Fort Stephenson.

At Fort Stephenson on August 1 and 2, 1813, a British officer Proctor was on his way to Sandusky River, with plans to destroy Fort Stephenson and then press on to capture Harrison and his stores. The order from Harrison, dated July 29, 1813 ordered Major Croghan to abandon the post, burn it, and retreat that night to Headquarters. Above all, defeat was not to be risked. This order was sent by interpreter John Connor and two friendly Indians who lost their way in the dark. George received the order the next day and called a council of officers, who determined not to abandon the fort. It was too late and perceived to be more dangerous to retreat in the open, in face of a superior force of the enemy.

George wrote to the General, "Sir Yours of yesterday (29th), 10 o'clock P.M. was received too late to be carried into execution. We have determined to maintain this place and by heavens we can."[6] This letter was perceived by Harrison as insubordination. He sent an escort of dragoons to relieve him. Major Croghan was arrested and brought before the General the next day. The Colonel stood his ground, by explaining he had sent a message as "bullying a one as possible the damned rascal Connor had failed to explain."[7] Harrison agreed for Major Croghan to return to his post, but if the British were discovered coming by water they would be coming with heavy artillery; to retreat, if not possible to defend the post to the last extremity.

There was joyous shouting when Major George Croghan returned to the fort. He started immediately strengthening the fort's position by working day and night. A ditch six feet deep and nine feet wide was dug surrounding the fort. Heavy logs were hoisted to the top of the palisades which could easily be pushed off to fall and crush any enemy that entered the ditch and attempted to advance.[8]

All the stores of munitions & supplies at the post were collected in one building that they might the more easily be destroyed if necessity required. The men prepared an abundant supply of cartridges. Rumors were thick that the Indians were on the warpath, and that British General Proctor was seeking to draw General Green Clary out of Ft. Meigs and fight in the open. A very big welcome would be given the enemy by the Major and his 160 men with one six-pounder canon. Major Croghan took time to write to a friend, "The enemy are not far distant. I expect an attack. I will defend this post to the last extremity. I have just sent away the women and children, with the sick of the garrison, that I may be able to act without encumbrance be satisfied. I shall, I hope, do my duty."

> The example set me by my Revolutionary kindred is before me.
> Let me die rather than prove unworthy of their name.[9]

On the afternoon of August 1, Proctor with 500 regulars and 700 to 800 Indians, appeared before Ft. Stephenson with gunboats. Major Croghan greeted him with a few shots from his single cannon. Colonel Elliott and Major Chambers, holding a flag came forward to demand the surrender of the fort. He was convinced it would be a bloody bath for those in the fort and was anxious to spare the effusion of blood. Major Croghan's answer was that the place would be defended and no force, however large, would induce him to surrender. The fire from the gun-boats in the river then opened on them and continued throughout the night. Major Croghan moved his lone six-pounder about changing places from point to point in the hope of deceiving and creating the impression that he had several cannons.

About 4 P.M. discovering the fire from all of the enemy's guns was concentrated against the northwestern angle of the fort, Major George Croghan became confident that their object was to make a breach and attempt to storm the works at that point. He therefore ordered as many men as could be employed for the purpose of strengthening that part; by securing it by means of bags of flour, sand.

Following Major Croghan's directions, the one gun was lifted up into a blockhouse and so placed as to rake that angle and the portion of the ditch leading to it. The port-hole was masked and the gun loaded with a double charge of leaden slugs. Major Croghan's inference was correct. The enemy attempted to make the assault about five o'clock under cover of the smoke from the battery. They were within twenty paces of the ditch before they were discovered; when they were checked for a moment by fierce musket firing from the fort. But they were quickly rallied by Colonel Short, who, springing over the outer works into the ditch, commanded his men to follow shouting, "give the damned Yankees no quarter." When the ditch was well filled the masked porthole was opened and the six-pounder was fired into the human mass, only thirty feet away, with appallingly fatal effect, while all the time the muskets of the fort never ceased to send their fire. The British were thrown into hopeless confusion, and all that could, fled fast as possible. Their loss was something like 150, including Colonel Short among the dead. Croghan's loss was one dead and seven slightly wounded.[10]

After the heavy fighting had ceased, Major Croghan knew the wounded British were in the ditch and not safe for their comrades to come near for moving them. Those inside the fort would have been shot if they had attempted to move them. It was known that a wounded soldier had a strong thirst for water.

It was said that the water was so contaminated in the Fort, men from the garrison had to leave by a back gate and bring water from a nearby spring. Colonel Croghan managed to have buckets of water passed to the wounded whose thirst always tormented those who had been hurt. Under the cover of darkness, a ditch was opened under the pickets through which many of them were taken into the fort and cared for. Major Croghan asked his men to use some of their clothing to care for the wounded. A week later, the Major's request for replacing their clothing was not answered.[11]

Colonel George Croghan, like his father Major William Croghan, as well as great uncle George Croghan and Sir William Johnson, had shown respect for the Indian. While fighting in the Revolutionary War, Major Croghan wrote about a deplorable act that Americans committed against the Indians after winning them to a religious faith. Americans then turned against them and slaughtered a whole Indian village which Major Croghan found barbarous and was unable to forget.[12]

At Fort Stephenson, about 3 o'clock in the morning of August 4th, Proctor and Indians abandoned the field and the campaign started back to Canada. Proctor's men had been defeated.

General Harrison sent his report to the Secretary of War with praise and high tribute to Croghan's gallantry.

> It will not be among the least of General Proctor's mortifications to know that he has been baffled by a youth who has just passed his twenty-first year. He is, however, a hero worthy of his gallant Uncle-General G.R. Clark, and I bless my good fortune in having first introduced this promising shoot of a distinguished family to the notice of the Government.[13]

All newspapers were full of his praise and his name was on all men's tongues. Lieutenant Colonel Croghan was assigned command of an expedition to recapture Mackinac Island and to break up British posts on Lake Huron, but it was the British naval superiority that decided the fate of this battle.

A month later, a victory occurred for Commodore Perry on Lake Erie. Early in October, William Henry Harrison, Isaac Shelby and the Kentuckians destroyed the British army and killed Tecumseh at the battle of the Thames and this ended the war.

It has been written, "The people of later generations can hardly realize what a hero it made of the young Kentuckian who commanded that gallant defense." The whole length and breadth of the land, Colonel Croghan's name was upon everybody's lips as "the hero of Fort Stephenson."[14]

CHAPTER FOUR

George and Serena

Very soon, George was in New Orleans fighting under General Jackson's command. Serena's Uncle Edward Livingston lived in New Orleans from 1803 to 1831 and had volunteered as an aid to General Andrew Jackson in War of 1812. Edward's son Lewis was in General Jackson's army in 1814–15 as an assistant engineer, with the rank of Captain and was in several battles. Serena's brother, John R. Livingston, Jr., as well as Henry Lee were fighting in New Orleans.[1]

Another person entered George's life while fighting in Jackson's army. John Wesley Jarvis was in New Orleans painting those historical full-length portraits of military leaders and naval heroes in 1812. Andrew Jackson's famous full-length portrait was painted by Jarvis at that time.[2]

After the War of 1812 ended, Henry Lee had returned to Virginia in 1816 and found himself in desperate need of money and in immediate danger of losing Stratford since his father had gone to debtor's prison. He chose to marry his young nineteen-year old neighbor Ann McCarty, who had inherited a fortune and large estate which enabled her new husband to continue his "country gentleman lifestyle." Unfortunately, Lee had some very offensive personal charges brought against him relating to involvement with his wife's young sister. He was then ostracized from society forever.[3]

Henry Lee had married Anne McCarty, March 29, 1817 and Serena married George Croghan two months later, May 1817.

In June 19, 1816, Major Croghan had written that son George would be home in five or six weeks. He had left New Orleans June 1816, and made a stopover at Locust Grove on his way to New York.[4]

Livingston and Armstrong descendants have always said, "George and Serena met at a ball in New York and it was General Armstrong who introduced his niece to Major Croghan."[5] July of 1816, George had just resigned from service and "happened to be visiting at Red Hook, New York, home of former Secretary of War Armstrong."[6] The former gener-

144

al seemed to be in seclusion after being blamed for the burning of Washington City and the White House. He retreated to his New York home Rokebe and his farm in Baltimore.

> The sarcastic Secretary of War, accused of treason, was driven away by what he called a village mob, and not suffered even to resign at Washington, and was forced by public indignation, to fly to Baltimore to do it.[7]

Serena's brother, John R. Livingston, Jr., her Uncle Edward Livingston, his son Lewis or other Livingston relatives were the ones most likely to have accompanied Colonel Croghan to New York after leaving New Orleans. One or all, could have been instrumental in bringing about the first meeting of George and Serena, and perhaps there were ulterior motives in bringing home the handsome hero to a young girl who had lost her love to another.

If Serena had any dreams of a future with Henry Lee, they were hopeless and shattered. Serena was also recovering from the death of her sister Angelica, a much greater loss than an affair of the heart. Reading women's diaries of the period, it is clear that getting a husband was uppermost in a young girl's mind and those around her seemed willing to assist in any way they could in this pursuit.

That winter, 1816 and 1817 while in New York, Jarvis painted the two portraits of Colonel George. Years later, Serena wrote that Colonel George was 26 years of age when these were painted.[8]

Before leaving New Orleans, George was recognized as a national hero, but there were others who were jealous. General Jackson was about to relieve George of his command and place Major Thomas S. Jesup, his future brother-in-law in his position. George had asked for a furlough before Jesup arrived, but received none.

Many years later, September 1827, Inspector General George Croghan, at Jefferson Barracks, after witnessing a certain event, wrote in his report of a similar event which occurred when he left the military.

George recalled the year 1816, which was a result of Indian agents having the same control over certain stores at posts as the military offices in command themselves have, and could at will obtain orders directing those commanding officers to erect quarters for them, "with soldiers perhaps already worn down by long continued fatigue in the preparation of their own."[9]

Among the last military orders which I received at New Orleans previous to my resignation in 1817 was one directing me to send a detachment from my command to some place in the Creek country to report to an Indian agent for whom it was to build houses; and forsooth, I received a severe written reprimand for not having sent him better men. In stating to you as in duty bound the feelings of the officers here in relation to the subject, I may be indulging my own personal ones too far.[10]

George had returned to Locust Grove soon after meeting Serena, but was back in New York that winter. While in New York, George wrote a letter to Jesup dated, February 27, 1817.

After the trial of Gen. Gaines and being disappointed in this, I then determined on remaining silent until [I] should have it in my power to tell you something, as to my future plans. This moment having arrived, I can now say to you 'that I have left the army, that I am shortly to be married and that I have resolved on establishing myself as a farmer in the neighborhood of Louisville and there to enjoy 'time cum digniti' (as the landowners have it). The young lady to whom I am engaged is the niece to Edward Livingston of New Orleans and is the eldest daughter of J.R. Livingston of this city [New York]. Her fame has probably reached you for she has been for sometime, as the 'fairest flower of the Hudson.' I can not state precisely the day on which I am to be received upon the respectable roll of Husband as it depends upon the return of the brother from Europe. Yet, the probabilities are in favor of it's taking place on or about the 1st of May (sooner if the brother arrives).[11]

The brother, John Robert Livingston, Jr. was a very important person to George and Serena. He had made a trip to France to return his sister Angelica's body for burial in New York. George and Serena waited for his return from Europe before the marriage took place. It appears that Thomas Sidney Jesup was another who George was anxious to have present for the wedding.

The marriage of George and Serena took place in New York, May 1817, probably at the Trinity Church where Serena's siblings and cousins had married or in the Dutch Reformed Church. She was twenty three years old and George was twenty six at the time.

Even though, Colonel George came equipped with courtly manners and a well refined education, it could have been intimidating to be suddenly thrust into the middle of the Hudson River Valley Society. This stronghold included Livingstons for miles around. George was a free spirit and probably would have preferred spending his time in the woods rather than in drawing rooms.

Soon after the wedding, the newlyweds traveled to Kentucky and a welcoming celebration would have taken place at Locust Grove. Meeting Colonel George and his bride was an event the Croghans and Clarks would have celebrated in grand Virginia style. Clarks, Gwathmeys, Thrustons, Andersons, Taylors, Prestons and many more would have put their best foot forward to meet the niece of Chancellor Robert and Edward Livingston of New York.

Cousin Richard Clough Anderson, Jr. wrote in his diary:

> June 24, 1817. I was in Louisville[.] Colo[nel George] Croghan &
> Lady from N York arrived at Majr. [William] Croghan's.[12]

In May of 1817, Major William Croghan had extended invitations to cousins and other relatives to visit Locust Grove. All the Croghan children were to be home that summer. William Clark wished very much to be there for the happy occasion, but sent his regrets.[13]

Very soon, Serena found herself witnessing events in Kentucky which she may have found strange. It was quite common for a delegation of Indian Chiefs to pay a formal call on General Clark where most often he would be found seated on the long north porch. Years later, recalling an Indian ceremony which took place on the back lawn at Locust Grove, George and Serena's great Granddaughter, Mrs. O'Meara of California said the family was expected to take part in the ceremony and all were seated in a circle.

> My great grandmother, a New York Livingston, was obliged to take
> her turn eating (I believe it was fish) when it came her turn, her husband whispered to her that she must eat or the Indians would be
> greatly offended if she did not.[14]

The newly married couple traveled to New Orleans after their stop at Locust Grove. They were now owners of a plantation in New Orleans, located near Edward Livingston's place. Serena brought this property to the marriage. However, they were back in Kentucky when George's uncle, General George Rogers Clark died, February 1818.

George and Serena wrote from Red Hook, New York, sometime in the summer of 1818 where their first child, Angelica was born. This baby daughter was named for Serena's sister who had died in Europe. The birth of their second Croghan child took place at Locust Grove in January 1819, while William Croghan, Jr. was in New Orleans overseeing George's plantation.

John Jacob Jarvis reappeared in Colonel George's life in winter of 1820 at Locust Grove. The portraitist was very fond, indeed, of his Southern friends, and he was no less lavishly entertained by them.[15] Jarvis and wife lived at George's boyhood home Locust Grove in Kentucky for several weeks. The previous year, Jarvis had married Mary Liscome of New York. During their stay at Locust Grove, Jarvis painted the portraits of Major Croghan and Lucy Clark Croghan which now grace the parlor walls at Locust Grove. This historic home and all it represents, is indebted to Angelica Croghan Wyatt's descendants for returning these treasures to their original home, Locust Grove.

An interesting evening took place at Locust Grove while Jarvis was visiting. Jarvis's granddaughter Mary Liscomb wrote about a dinner at Locust Grove during the extended stay. About ten people were present and several rounds of toasts had been given when George Croghan asked for a paper and pen before he made a toast in which he gave Jarvis's new bride (second wife) a lot between Locust Grove and the river near downtown Louisville. The granddaughter always regretted the loss of the deed which was destroyed in the burning of Richmond, Virginia during the Civil War. Today, this plot of land given to Jarvis is near Locust Grove and retains street names, Jarvis Lane, Jarvis Way, and Jarvis Woods.[16] There are indications that Serena was not present at Locust Grove during this stay.

Orlando Brown, son of the senator from Kentucky, wrote in 1820, "Col. Croghan and lady passed [through Princeton, New Jersey] a few weeks ago. He looks a good deal 'hen pecked.' He intends returning to Ky. in a short time leaving his amiable sweet tempered wife at her fathers."[17]

It was said that the Livingston daughters (Serena's aunts) were well versed in politics and literature and had contributed to the earnest discussions in the drawing room.[18] One can imagine, during Serena's life, she voiced her opinion on many subjects. Perhaps she didn't fit into the passive role which was expected of Southern women. Serena was well educated and had spent her young years in the presence of her Livingston aunts and was exposed to conversation which was not given to small talk.

CHAPTER FIVE

New Orleans

In 1824, President Jackson appointed George postmaster of Dubougs Post Office, New Orleans. Serena's family was not pleased and wanted George to settle down. There were two positions under consideration for George which the family thought would be much more suitable. John Croghan agreed whole heartily with the Livingstons. General Jesup received a letter from John, "I should have been pleased if he could have exchanged his office for one less lucrative in a different state. I dislike the idea of his locating himself permanently in that vile [illegible] of dissipation, New Orleans."[1]

There was always someone ready and waiting to separate the planter from money and his plantation itself. The newcomers to New Orleans were in surroundings which were hard to ignore, especially for George who had no family around at the time.

This willful decision made by George to settle in New Orleans was the beginning of his financial problems, from which he never recovered. The pay for a postmaster or an Inspector General was never quite adequate for taking care of the Colonel's needs, much less a family. Serena would not live in New Orleans therefore, she and children were back with the Livingstons in New York. Before leaving the post office to start his duties as Inspector General, he was faced with an unpaid debt to the post office for improvements made. He fell short in his settlement of more than $11,000. He had calculated on remaining as postmaster of New Orleans for many years and had used some of the money belonging to the post office in his private accounts expecting to repay this money later. At the same time, George had mortgaged his Louisiana and Louisville property for $62,000 to The Life and Fire Company.[2]

Colonel Croghan was not the only one to get caught up in the reckless buying and selling fever when he sold his plantation for houses in town.

After the War of 1812, the great migration west reached its peak. Clark cousins and Uncles were buying land for "get rich" dreams. Even before the war ended, Edmund Clark had moved to Kentucky where he and Jonathan were very much into land speculation.

Western Kentucky frontier was open for new settlers. Many who were into land speculation at that time were overextending themselves in ability to pay. George and cousins lived as though the use of money was always there for the asking. In the early 1820s, his cousins especially, Richard Anderson, Jr. and John Gwathmey had played the game even more heavily.

Richard Clough Anderson, Jr. was facing financial ruin and a decision was made to accept a U.S. Government appointment as Foreign Minister to Colombia, in order to save him and his family from ruin. However, his wife Ann, paid the price and died in that unhealthy climate where others had refused to go because of the unhealthy conditions.[3] George's Uncle Richard Clough Anderson, Sr., also was on the brink of losing everything he owned, until Richard, Jr. stepped in and gave more than he could spare to save his father's finances, even at the risk of financial ruin for his own family. While in Colombia, Anderson, Jr., received notice that the bank had placed claim on his own Kentucky property for his father's debts.[4]

Another cousin, John Gwathmey, was a business man about town. In Louisville, he was owner of the popular Indian Queen Hotel and other business pursuits until he was so heavily in debt due to land speculation he was facing ruin even in 1818. Cousin Richard Anderson, Jr. wrote in his diary, "John Gwathmey has at last failed, that is his notes in Bank have been protested. It is not known whether his property will be sufficient, if it is not, his friends & relatives will be very much injured by him."[5] No one would co-sign a note for him. He left Kentucky and died penniless in New Orleans, 1824. Two years later, Richard Anderson, Jr. died in Colombia. He was still trying to pay his father's debts. George's uncle Colonel Anderson died in 1828, leaving his whole family with no other choice but to fold up at Soldiers Retreat and move to Cincinnati, Ohio.

Croghan's neighbor Richard Taylor's (Zachary's father) home Springfield had to be sold for debts when he died in 1829.

In George's case with the post office, as fate would have it, the company that held the mortgages on George's properties went under. Previous to the company's failure, the holders became indebted to Jacob [illegible] and assigned to him a number of securities and properties, among them were the New Orleans & other properties belonging to George.

As a result of George's indebtedness, the Croghan Family were very alarmed and put time and energy in finding a solution rather than suffer a great financial loss.

Philadelphia, December 7 [1826], Serena wrote to Lucy Croghan, Locust Grove, Louisville, Kentucky regarding a recent injury George had at the time of this financial crisis.

> Genl Jesup wrote you some days ago my dear Mother an account of Col. Croghan's illness. I am anxious to relieve your minds, he is a great deal better, and were it not for the wounds in his arms would be perfectly well. Dr. Physic has been here this morning, and has pronouncd that his recovery will not be very tedious, Mr. James Biddle whose attentions have been unceasing obtained letters from the Physicians, which he immediately forwarded to the Secretary of War, & to such Gentlemen at Washington, whom he thought might prevent any rumours that might be in circulation, from being injurious to him.
>
> The physicians will not yet permit him to remove, until his right arm can be put in a case,
>
> Your sincerely attached
> Daughter
> Serena [Livingston Croghan]
>
> Col. Croghan desires me to say to William that he will have his affairs with the Life & Fire Insurance Company settled and that I have already written to Papa [illegible] investigate the matter & to have a suit in Chancery brought.[6]

William traveled to Philadelphia to approach Serena's father, John R. Livingston for a $6,000 loan for George. The loan evidently didn't come through and the bank would not approve a loan. Serena's brother, John R. Livingston, Jr., an attorney went to New Orleans to assess a perfect account of the whole transaction and with the opinion of Counsel on the proper course to pursue. He stopped off for a weekend at Locust Grove on his way to Louisiana. Afterwards, William received a letter from John Livingston, "I must by you to remember me to the many kind friends I left at Locust Grove whose attention to me when there I shall never forget."[7]

Following this unfortunate event, George's brother, Nicholas died, leaving land for George in his will. This sale of land saved him from a miserable situation.

May 1828, George wrote to Jesup from New York, "My family, are now in the country at Mr. Liv. where, they will remain until the fall."[8]

For some time, reports were coming to George about William Harrison starting a campaign to discredit and belittle George's heroic defense of Fort Stephenson. This campaign against Colonel George's bravery started immediately after William Harrison was elected to the Senate in 1825 where he became chairman of the Committee on Military Affairs.[9] Harrison was then in daily contact with Quartermaster General Jesup. The criticism of Col. George's military record never let up. Those around Harrison were the ones who made the derogatory remarks as if they were coming from someone other than himself. The remarks were passed on to George from different sources.[10] Harrison kept this up because he, himself, in his senatorial campaign was criticized for misrepresenting some of his war records and his actions at the time. Even the women of Chillicothe, Ohio, had presented the Colonel a sword for his brave actions and as a result saved their town from being burned and taken by the British. The women also sent Harrison a petticoat.[11]

Colonel Croghan's leadership in the defense of Fort Stephenson had a great impact on the nation at that time. Harrison's leadership had met with disaster, one after another until the Siege of Fort Meigs in early May, 1813. He fought the British off successfully, but the cost of lives left nothing to celebrate. Perhaps, Harrison was jealous and resentful of all the national praise and hero worship given to Croghan.

One writer says Croghan put too much into his relentless fight to defend his heroic deeds, by fighting back instead of ignoring Harrison's attacks. On the contrary, records show that Harrison was the one who went to great lengths in dismissing Croghan's heroic battle as nothing.[12]

Freeman Cleaves wrote in Old Tippecanoe, "Young Croghan had won laurels at Tippecanoe and Fort Meigs but Harrison apparently spoiled the Major's chances of winning additional honors by giving him instructions similar to those left with Colonel Samuel Wells in June. Should the British approach in force with cannon, Harrison ordered that Croghan was to burn the fort and retreat. If only Indians appeared he was to remain until relieved."[13] Above all else, the General emphasized, defeat must not be risked.

Colonel Croghan had shown true leadership in the defense of Fort Stephenson and was recognized with honor. A much needed hero and victory had not appeared on the scene until the twenty one year old boy had fought off the British and won. One year later, America suffered a big defeat when the White House was burned by the British.

Again, Colonel Croghan challenged Harrison's attacks and fought back. Letters went back and forth between Croghan and Harrison in which Croghan demanded Harrison's denial or he would expose the letters sent to him by Harrison in which he praised Croghan for his heroism.

General Harrison had held Colonel Croghan in high esteem and had written a letter to the Secretary of War early in 1812. Harrison had recommended this young man should be an army captain. He had been a soldier,

> who certainly conducted himself in the most exemplary manner, in every station in which he served. Although brought up with all the delicacy that is common in [an] affluent family, he performed all the duties of a private Dragoon, with the most zealous assiduity; and was always amongst the foremost to volunteer his services in the fatigue parties, that were called on to work on the New Post, on which he laboured many days although his Corps was exempted from the fatigue detail. I can also assert that he possesses all the courage and fire which are so necessary to form a good officer.[14]

After so many others got into the heated dispute in 1825, this prompted a request from Harrison to recruit mutual friends to act as a board of honor in settling this impasse. General Jesup, Henry Clay and Colonel Josiah Stoddard Johnston brought about an agreement. It's easy to conclude that General Jesup (who was always jealous of brother-in-law, George's fame) was behind Harrison's belittling of Colonel George in the first place and to stop Colonel George from releasing the letter, General Jesup was the one to suggest a board of honor, himself in charge of the board. George answered Jesup's proposition made to him.

New Orleans, 19th Nov. 1825

> Genl. Harrison & I have been for some time in correspondence in relation to the Sandusky affair, his last letter I have not yet replied to first because I have been too much occupied since its receipt and again because I have not as yet fully decided upon the pro-

priety of agreeing to the proposition which it contains which is to refer the whole matter to some of our mutual friends, say to yourself or another chosen by you. I believe that I will accede to the Genl.s terms, but cannot answer positively. The more I think of all I did for Genl.Harrison of his subsequent [illegible] than subject of me, of my consequent mortification in being passed upon by Congress as unworthy of an expression of its approbation, the more am I urged to place everything before the world short it may be [illegible] whether or not I have cause of complaint. Advise my Dr. Sir. G. Croghan.[15]

Colonel Croghan found it necessary to send a second letter to General Jesup.

New Orleans 20th Decr.1825. Dear Genl....I hereby inclose to you for the information of the board a copy of the correspondence between the Genl. [Harrison] and myself since July 1818. ... My letters to Ben Harrison contain the sum of all my charges or complaints against him, and may be considered as all that I shall offer unless he shall contradict any of their material statement [.]Yours truly G. Croghan.[16]

Harrison had just been elected U.S. Senator and his reward was Chairman of the Committee on Military affairs. In the fall of 1828, he left the Senate to become the American minister to Colombia. His success as a diplomat didn't last long. He encouraged a revolution to overthrow Simon Bolivar and President Jackson relieved him of his post.[17]

In 1835, Senator George M. Bibb of Kentucky introduced a resolution in the Senate and spoke of the defense of Fort Stephenson as "the cause of saving Ohio and the adjoining states from invasion, from desolation, from plunder, and from bloodshed. The Western country was saved from the hostile and destructive incursion of the British and Indians."[18] Congress authorized the striking of a gold medal in Col. George Croghan's honor. On February 8, 1815, swords of honor were awarded to Colonel George Croghan, Major General Jackson, General Harrison, Governor Shelby of Kentucky and five others for military virtues.

Years later, Harrison was a candidate for president and again, he began to attack Colonel George Croghan. Colonel Richard M. Johnson of Kentucky (Harrison's right-hand man) suggested a history be written about the War of 1812 with Harrison's careful approval of everything that was written before publication.[19] It was all favorable to Harrison, Shelby, Johnson and others, but not a single mention of Colonel Croghan's name. Military figures, other than Croghan protested and disputed this version of the war and found it necessary to defend themselves for their bravery. Even General Andrew Jackson replied to this publication for criticism printed about others.[20]

1840, Harrison was on his "log cabin campaign," running for United States President. His public image had been so inflated by political rhetoricians, a certain Washington D.C. newspaper editor could not tolerate this any longer. Francis Preston Blair was a Kentucky schoolmate of the Croghan boys, who came to D.C. as President Andrew Jackson's hand-picked editor for the only newspaper, *The Globe*, in the nation's capital. Blair wrote a scathing piece about the coward, Harrison, but not before President Jackson got into the fray and defended Croghan loud and clear. Editor of *The Globe*, James Blair, minced no words when it came to explaining the difference between a coward and a hero.

> Harrison's military career, however, was another matter, and Blair bitterly resented the equating of Harrison with Jackson. *The Globe* insisted that Harrison's mistakes at Tippecanoe had caused many unnecessary deaths. Blair's beloved uncle George Madison, the erstwhile short-lived governor of Kentucky, had blamed Harrison for the American defeat at the Raisin River, and Blair repeated this charge in detail. There had also been ill-feeling between the famous scout George Croghan and Harrison, and Blair appealed to the Irish vote by emphasizing this. At Fort Meigs, wrote Blair, Colonel Croghan had saved Harrison's camp from destruction, but had received from Harrison only 'injustice and ingratitude.' The ladies of Chillicothe, Ohio, however, had not been deceived. According to the Globe, these wise women had prepared a petticoat to present to General Harrison at the time that a sword was presented to Colonel Croghan. And finally, Blair insisted, 'Harrison's one major victory, the Battle of the Thames, had been won primarily through the advice and battlefield heroism of Colonel Richard M. Johnson, said *The Globe*, who had 'plenty of wounds on his body, but not a single certificate of bravery in his pockets!' In contrast, Harrison had 'plenty of certificates

of bravery in his pocket, but not a single wound in his body! Mark the difference between a real hero and a sham hero![21]

William Henry Harrison was elected President of the United States in 1841 but died one month later from pneumonia. Strangely enough, "Harrison, on his death bed, he sank into a stupor, his mind wandering in delirium. Colonel George Croghan, who came to the bedside at Harrison's request, testified to broken sentences; 'I cannot stand it. I cannot bear this. Don't trouble me.[22]

Harrison had lost four grown sons, Symmes, William Jr., Carter and Benjamin. William Jr., intemperate to the last, died February 6, 1838, and on August 12, 1839, Carter "was suddenly and mysteriously" taken away.[23] Harrison, while campaigning, had called on John O'Fallon, after no contact for years, and asked for help in committing his son in some institution in Missouri.

CHAPTER SIX

Inspector General on the Frontier

George Croghan, Inspector General in the United States Army, was a military leader who was away from his family and friends for long periods of time. The western rivers marked the limits of the frontier along the Mississippi River from New Orleans to Minnesota and Wisconsin. Each post was cut off from the world and isolated into itself in the vast wilderness, but the soldiers made an outstanding contribution to western development. Colonel Croghan's tours started from early spring until fall and these tours continued for twenty years. He was concerned for the safety of the army posts in any attack from England, France or Spain. He looked out for the soldiers and wanted the best for them, but also, he respected the Indians and was concerned that their lands and hunting grounds would not be destroyed. Small settlements begin to appear around the posts and over the years grew into small towns. Colonel George Croghan, like his father Major William Croghan and Uncle George Croghan of Pittsburgh, contributed to the "rise of the new West."

Most of Serena's years of marriage to George were spent with other family members or living near them. She was never settled for long and moved back and forth between New York, Washington City, Baltimore, Philadelphia and Louisville. This unsettled life was not uncommon for military wives, but planning and packing the necessary supplies for her young children, servants and fine clothing for herself, must have been an enormous task when preparing for travel. Most of her traveling was by water, but there were times when she had no choice.

After the New Orleans debacle and loss of property, life was never the same for George and Serena. He seemed to be in financial stress the rest of his life. Army pay for Inspector General was not enough to support himself while traveling on the frontier and supporting a family back East. This was a stressful unexpected way of life for Serena and George. Their marriage was a subject of many debates among Locust Grove docents of long ago. Did George's drinking turn Serena into a shrew or did the "taming of the shrew" drive George to drink. After George soon turned to bouts of intoxication, it wasn't a smooth happy marriage and in turn affected the lives of other family members.

Colonel George was overwhelmed with stressful situations from all sides. He probably dreaded going home to Serena's badgering, but she also, had good cause. His need for money was a constant worry. Harrison's attacks would have been difficult for anyone to live with.

When George, Inspector General of the U.S. Army, was traveling from one army post to another, in remote areas, drinking was prevalent among the soldiers in the army. Even in War of 1812, whiskey was first on the list of needed supplies. Those remote areas, far from civilization, did not lend themselves to a disciplined lifestyle. The daily existence in the undeveloped frontiers was reflected in soldier's letters of how much the society, family, and friends were missed. George comes through as a loving father and rushed to his family when there was sickness. There is no indication of him ever being a womanizer.

As time passed, it became obvious that Serena had lost respect for her husband. He must have been beaten down by her scolding in the presence of his children, Livingstons, Croghans and others.

August 1828, William Lindsay, Colonel USA, New London, Connecticut wrote to Jesup about George, "last stages of intoxication."[1] How did a bright future at one time turn to this sad state of affairs? George had intelligence, personality and was recognized nationally as a hero with new friends and acquaintances. Caught up in the moment of fame and glory, the belle from the "fairy land" world on the Hudson River married the soldier from Kentucky who loved to ride through the woods and hunt. Their worlds had been far apart.

In those days, entering into marriage was a lifetime commitment. It might have proven wise for those contemplating marriage to have had a long courtship, giving each an opportunity to know each other. As one young lady told one who wished to marry her, "No, because, you will find that I am a cross, ill-contrived piece of stuff." Many young men should have been so lucky to have heard this statement before marriage. She continued, "says I, women are bad, but men are so much worse that it was a wonder if they agreed."

George and Serena increasingly had their problems. Family letters reveal the unhappy situation for all those involved as Colonel and Serena's problems increased and Serena began to come under criticism from family members. Her circumstances were different from others in the Livingston or Croghan families and there were times when it was doubtful whether a monthly payment for household expenses would be coming through due to George's drinking. Evidently, George's monthly pay checks were to go to Serena.

Serena and children were living in Barrytown, New York, 1830, when John Croghan arrived in New York to escort George on his tour for disciplinary reasons. Charles Croghan was to stay with Serena and children while George and John were on tour. Jesup received a letter from George, "I have been a little lazy or rather pretty much occupied in parading the struts with the Doct, to whom I have been showing the lions of the place."[2]

While on tour, George was suddenly called home due to the illness and deaths of his children. "I am now in attendance upon my sick family. Two of my children have been wrested from me within the last three weeks and of the remaining two, one [Angelica] is scarcely yet convalescent and the other [St George] is now stretched upon his bed, ill of the very fever which has already visited us so fatally."[3]

April 9, 1835, George and family had returned and were living in Georgetown, Washington City. He could not resist the temptation of drinking with bad companions who were always waiting upon learning of his arrival in town. When he was completely intoxicated, they took his money. While living in Georgetown, General Jesup received a letter from Serena:

> I am convinced that Col. Croghan has been induced to remain out all night by the Mr. Bird of whom I spoke to you yesterday. A hack man came here this morning to tell me that he was sent by Col. Croghan to say that business before the Committees had kept him out all night and that I must send the Carriage to the Capitol at 2 o'clock for him. Nathan [servant] met the same man & insisted on his showing him where Col. Croghan was which he refused. He discovered him however in a miserable little Shop or Tavern on the Avenue. The people of the house did not know Col. Croghan but said that he was brought there by two gentlemen at twelve o'clock last night. (one of them Nathan ascertained to be Mr. Bird). They paid for his breakfast and night lodging for he had not a cent altho' he had been seen early in the evening with gold and bank notes. I

have myself no doubt that this Mr. Bird who urged him not to return home last evening ... has induced him to accompany him to some Gambling House where he has taken advantage of his situation & plundered him. I only fear that he may have induced him to sign some note to a large amount. Can you ascertain O Major Cross if he drew yesterday the quarter due for the next month. His pay was drawn & given to me the day before yesterday. He is now in a terrible situation & can not speak. You will oblige me by requesting the pay master not to pay him any money for the next month.[4]

The following years, General Jackson was removing the Indians west of the Mississippi. Ann Jesup and children were living at Locust Grove while General Jesup was chasing Oseala in Florida. Not only was Ann grieving over the death of her mother Lucy, in 1838, it seems her husband, General Jesup unloaded all his troubles concerning brother George on his wife, Ann. Reports of George's behavior would be brought to Quartermaster General Jesup's attention. The following letter is from Ann in response to Jesup's never-ending complaints about George.

But had my poor lost brother had a kinder and more affectionate wife he would have been a different man, he never talks of all her unkind treatments to him, my mother has shed tears while telling me of it, you will hear so much of his bad conduct that it will be enough to make you cast him off entirely, but my dear husband he is my brother, one whom I was once proud to call brother and although now disgraced by his habits, still I have a sisters affection for him. Is it not strange that he will be here for three or four months and at our house as many or more and never gets intoxicated?[5]

Lucy Clark Croghan made a will in 1830, giving everything she owned to George's children and until they became of age, daughter-in-law Serena was to receive all rents from her houses in Louisville. Dr. John had not informed Serena of the contents in Lucy's will. When Serena inquired of John, she received information that his mother had made him administrator of her estate until the children became of age. John also told her it would take a few years before the children would receive any rents money because this money was owed to others.[6] Did the "others" mean Dr. John and General Jesup?

Serena wasn't the only one who received unfavorable news from John concerning the administering of his mother's estate. Once again, Dr. John was determined that George Hancock was not being honest with him. This time, it involved 300 acres of lands purchased by Hancock from Charles before his death. In answer to a letter received from General Jesup, George Hancock said the Doctor was mistaken in the claim. He had never sold the Doctor anything but the remainder in the Locust Grove property and sent a copy of the agreement between Hancock and Charles. He had purchased 300 acres of Charles for $6,000 payable on long time and Pirtle and Anderson had drawn the contract. A will of Charles was then produced as his last will, giving the same property to Colonel Croghan's children.[7]

Hancock, determining to have no conflict with the Colonel's children, when he could legally have maintained his contract, destroyed the contract for the purchase.

Charles Croghan's twin Nicholas died in 1824. In Nicholas' will, he left niece Angelica Croghan, 830 acres in Warren County of Big Barren River. To his nephew, [St.] George Croghan, 1505 acres Big Barren River near Bowling Green. Sister, Ann Croghan Jesup and brothers John and Charles were included in his will.[8] The youngest child of George and Serena, Serena Livingston Croghan [Rodgers] was not born until 1834, after the will was written, so therefore would receive nothing.

The other Croghan twin, Charles, died in 1832, leaving Sister Ann Jesup property and money due from Underwood and brother-in-law Mr. Hancock. To children of Brother Colonel Croghan he left real estate including farm, six miles above Louisville.[9] Again, the young Croghan sister (Serena Croghan) would not be included in Uncle Charles' will.

Grandmother Lucy Clark Croghan died in 1838, but her will was written January 12, 1830. Listed in the will was Grandson John Croghan, who died two years later in 1832, leaving all proceeds going only to Angelica and St. George, the other two living children of George and Serena. Serena Croghan Rodgers received nothing since her name was not on the will.[10]

In September 1840, Jesup was writing from Washington to Dr. John regarding a bout with George. As usual Ann got the brunt of his anger.

My Dear Sir,

The treatment of Colonel Croghan to me, and the serious embarrassments I have experienced in consequence, has been a source of constant mortification and distress to her. She was aware of the importance to me at the last session of Congress, of seeing and associating freely with the members, and knows the injury I have sustained by not having at it in my power to do so; and because her brother [George] had deprived me of the means, she seems constantly to blame and reflect upon herself, as if she were answerable for his conduct. He was, and is, bound by every principle of honor & honesty, to pay me the [illegible] I [illegible] & paid for him, from his pay accounts this treatment I shall not very soon either forget or forgive. I had promised Lucy Ann to take her to the best School in Philadelphia, to finish her education, but not having the means to do so, she has lost the best year of her life. My children if educated at all, must be educated from my own means. You are no more accountable for your brother's debts than we are.

Th S. Jesup.[11]

No doubt, those were trying times for Serena and children. Their private lives had become public. Her only solution for saving herself and children from further stress would require courage and spirit. She wrote letters to Dr. John and William Croghan, informing them that she was contemplating divorce. William sent a letter, urging her to reject this idea, explaining to her that such a drastic step would bring shame and disgrace on both her and the children.[12]

It is conceivable that the Livingston family may have encouraged Serena to take action, even though there were unpleasant consequences. Serena's brother John Livingston was there at her side assisting and advising her in protecting her possessions by taking legal action. June 1840, an indenture was made between George Croghan of the U.S. Army, of the first part and John R. Livingston, Jr. of the second part whereas witnessed the party of the first part, "received and obtained by his wife Serena E. Croghan who was Serena Livingston of New York, a considerable estate which he, the said George Croghan hath used and employed, in paying his debts." The document also includes the considerable landed estate in Kentucky, "as appears by the will," was left by the children's Uncle Charles, depriving them of rents. "Witness the hand and Seal of the said George Croghan the 19th June 1840."[13]

The indenture was not filed and recorded in the Jefferson County Clerks Office until December 1844, when Serena left George and moved to Philadelphia, just before his Congressional Court Of Inquiry took place.

Serena was desperate and the choice was hers. This was her only means of stopping George from selling her possessions. Again, her brother advised and acted as her attorney. The matter of "selling her possessions" had touched other family member's lives, as pertaining to a letter received by Ann from William Croghan, August 9, 1835 concerning a missing watch. She or General Jesup had learned of George pawning a family watch which once belonged to William, but was very surprised and very unsettled about the news since the watch was a treasured valuable possession.

He was glad that the mystery of its fate had cleared up. First, Serena had been in great error in supposing he had lost a watch of George's. When Dr. John returned from Europe wearing the watch that Charles once had, William had made attempts to pay its value to have the watch back and had even went so far as to interest his Mother in his behalf in the matter, "there was a sort of hugga mugga with the Dr. in the affair, I could not comprehend." Where the watch was he could not find out and a deaf ear had been given to all his appeals for information on the subject. He wrote, "You may inquire why my anxiety to have it? I will tell you, but remember on no terms now would own it".[14]

Before his marriage that watch had been admired by his poor dear Mary and after their marriage, he had presented it to her. She requested William to take charge of it for her and she would direct him how to dispose of it. After the birth of their son, she reminded him of the present and wished it for the son. On the death of poor Nicholas [Charles Croghan's twin] he had felt a brother's solicitude for Charles. When William and Charles were requested to make a journey on his account, they were off for Washington D.C. Charles had wished for a watch and wanted to sell a field of corn which he wished sold and proceeds invested in a watch. To soothe the poor fellow, William told him that his watch, he might have on conditions he would not part with it unless to me and he consented. "My wife was hurt when she heard I had parted with it, but when I explained the circumstance [she] was satisfied & often told Charles he must keep it for my poor little Will. I wished it as a relic, precious to me from its history & associations—but now it has been polluted by the pawn brokers touch & trust I may never see it again. Yr aff brother William"[15]

One hundred and thirty three years after the above letter was written, an interesting discovery was made while Virginia Wheaton [George and Serena's great, great, granddaughter] was searching through stored boxes left by her mother. Virginia wrote to Mrs. Emmy Smith of Louisville, Kentucky, "Mother died on the 18th [1968]. We have kept her apt. but the thing I did come across was a watch in an old fashioned box with a note saying it had belonged to George Croghan. Opening the box I saw the lovely old gold "works" for a watch but the watch case had gone. I have it in Honolulu and shall send it on to you."[16]

The missing case, itself, must have had some significant value, such as an engraving of a famous maker and had belonged to a well known Revolutionary patriot. Charles Willson Peale was a successful watchmaker. These watches were engraved with his signature and were very much prized and valuable. Very few were made. George Washington received one as a gift from Peale. Thomas Jefferson and Peale were very old friends and Jefferson was given a watch made by Peale.[17] William Croghan made reference to the watch as, "relic, precious to me from its history and associations."

The true story of the watch has gone to the graves with those who were there, but regardless, the watch "works" ended up with Serena.

Was George Croghan's old drinking buddy "friend" P.T. Barnum of Washington D.C., involved in the mystery of the watch? Barnum was well known for his scheming, conniving ways. He never let up on anyone who possessed a relic which had historical value. His *American Museum* in New York was filled with many Revolutionary War pieces which he had gotten by suspect means. All these were destroyed when the museum burned to the ground in the 1860s, just before the Civil War.

An unfortunate incident took place in St. Louis January 6, 1841, during one of George's escapades. George had just boarded a boat after borrowing money to pay his passage. The boat sank and George was being arrested. George Hancock and the many friends (especially soldiers) who were loyal to George, came to his rescue and saved him. P.T. Barnum was in St. Louis at the time and had been harassing the Colonel for money owed him. In the middle of the dispute, Hancock attempted to pay Barnum the money, but Colonel George objected vehemently, insisting that Barnum was not to receive any money. For some reason, Barnum was not to be paid. It must have been a loud ugly scene. Hancock wrote to John, imploring him to take steps for George's resignation from military service and return home to Locust Grove.[18]

Serena had secured a legal separation and George had helped his family move to Philadelphia, but not before he had to hand over money from some source for his families' needs when moving to Philadelphia. After his family was settled in Philadelphia, Colonel George was living with John at Locust Grove and attending Temperance Society meetings. John wrote to nephew William Jesup, dated January 16, 1841, "Your uncle George and Zachary Taylor who is now here, desire to be remembered to all of you."[19] George had joined the Temperance Society and had been attending meetings, due to General Taylor's love for his old friend Colonel George. The Temperance Society Chapters at Military Forts were started for the benefit of so many drunken soldiers. George was staying sober, but in need of money and searching for a loan from some source.

For two months Dr. John had noticed something rendered George unhappy, even though he had acted in the most exemplary manner. George had told him the cause of his anxiety just before the Doctor received letters confirming at different times and places, he had fits of intemperance, involved himself and drawn for double pay. This had been officially made known and he was miserable and unable to raise the necessary sum to pay the accounts. He was more thoughtful and more penitent than John had ever known him to be and wrote, "If he can but once get free from the shameful embarrassment which his vile debaucheries have occasioned, that he will soon be what he ought long to have been."[20]

William had also, learned that the Colonel had drawn double pay for January, February and April. To save the Colonel, he sent one half and the Doctor paid the other half to save the reputation of one's family and couldn't hesitate making an effort to save him.

Dr. John was pleased to let Jesup know about his belief in George's changed behavior, "I made known to you that he had become a member of the "Temperance Society," and I candidly believe he will forever be true to his pledge. He remarked yesterday to me that he would be comparatively happy if he could be relieved of certain debts contracted (no doubt) during his—of intemperance. It is extraordinary, under the circumstances, the strong hold he still has upon the affections of the people."[21]

There was to be a Court of Inquiry in Washington D.C. March 22, 1845. George was in Washington. "I passed last evening with the President [Polk] & spoke about my anxiety to get away. He promised in the most friendly way to call the Secy of wars attention to this subject & immediately. I had a long conversation with him in relation to incidents of the last war, Genl. Harrison, Sandusky, &c which seemed to interest him greatly. I will return to Ky as soon as I am permitted, of this rest assured, for I am tired enough of this place."[22]

PROCEEDINGS OF A COURT OF INQUIRY
Held at Washington, D.C.
12th of May, 1845
In Obedience To The Following Order
WAR DEPARTMENT
Adjutant Generals Office

The President directs that a Court of Inquiry be instituted to examine in the nature of the transaction, accusation or imputation against *Colonel George Croghan,* inspector General, U.S. Army, of drawing his pay and allowances, twice for the same months of January, Febrary, March, April, November and December of 1844, upon double sets of Pay accounts duly signed and certified by him. The court will report the facts, and give its opinion on the merits of the case.

W.L. Marcy,
Secretary of War
General Orders # 16 EE77
Court of Inquiry
Washington May 12, 1845

Colonel George Croghan defended himself before the Court. The six page report he read from was Respectfully Submitted to the Secretary of War, W.L.Marcy. George was brilliant in his own defense. He was under great stress at the time. In January, he had just moved his family to Philadelphia and Serena had been awarded a legal separation. George made reference to this:

> ... Towards the close of the month of April 1844—being on a tour of official inspection duty. I arrived at New Orleans from Fort Pike. Having occasion for the use of money for certain matters of private business, and being on terms of the most friendly intimacy with Paymaster Denny, I applied to him to furnish me with such funds as I might require and offered to place in his hands pay accounts to cover the amount which I might have occasion for. At the time I was laboring under the influence of a powerful dose of laudanum, a medicine which by the advice of my physician I was compelled occasionally to use in consequence of violent attacks of neuralgia.

It was slow and hard for his mail to catch up with him on the frontier. Colonel Croghan admitted some difficulty in keeping his notes together and frequently was delayed in completing his final draft because trunks containing his inspection notes had not yet caught up with him in his travels. This author found in his old papers of reports one envelope with three stamped postal marks on the front. This letter had been forwarded four times.

> To: Col. Geo. Croghan, Inspect Genl. U.S.A.
> Sep 6—*New Orleans*
> Sep 14—Marked forward to *Washington*
> Sep 22—Marked forward to *Louisville, Ky*
> Found written at the bottom, forward to *Mammoth Cave*.

After five days of hearings, the following opinion was given.

OPINION

The court after deliberate examination of all the circumstances of the case of Inspector General Colonel Croghan, is of the opinion that the facts established, will not warrant the imputation of deliberate or intentional fraud upon the government. But the carelessness of Colonel Croghan and his apparent disregard of consequences in his transactions with Paymaster Denny are most reprehensible.

The accounts were given to Paymaster Denny, about the 1st of May 1844, and although Colonel Croghan, by his own admission, was in doubt, at the time or soon after, whether the same months had not been paid for previously, it does not appear that he took any steps to satisfy himself of that fact, or to make any arrangement with paymaster Denny up to the time of the receipt of Paymaster Denny's letter, which was about or after the 30th of September 1844. Colonel Croghan explains this as being on his understanding of it as private arrangement or transaction and that the accounts were given to secure.[23]

Like his Uncle George Clark, drinking was not the whole story of their lives. Making and keeping the frontier safe for new settlements was never brought to the public's attention. No one praised him or recognized his accomplishments on the frontier. The frontier soldier looked up to Colonel Croghan and had an attachment and love for him which surprised Dr. John and brother-in-law George Hancock.

He was a very intelligent and capable man who should have held the position of his brother-in-law, General Thomas Jesup, Quartermaster of U.S. Army. Jesup benefited by using Croghan's reports sent back to Washington from his inspections on the frontier. Those closest to Quartermaster General Jesup had always been his war comrades such as William Harrison and others, who insisted on playing down Colonel Croghan's bravery at Sandusky. Harrison was one who had a "scandal machine" when running for office. It was said, Washington and Jefferson also had so called "scandal machines." Jefferson was paying Calendar to slander Adams, which seemed to have backfired on Jefferson in later years. Washington listened to gossip and made notes at dinners.[24]

John O'Fallon was one who recognized the true Jesup. Upon O'Fallon's cousin Ann Croghan's marriage, he felt she had made the wrong choice for husband and should have chosen Biddle. Years later, a letter was written by O'Fallon in which he says, "Jesup is writing to The Globe making slanderous remarks about Gen. Scott's name in order to get General Jackson's attention."[25] The scheme was successful and as a result, General Scott was recalled from Command while fighting the Indians in the Everglades of Florida and Jesup replaced him.

Jesup was keenly aware of Croghan's intelligence and leadership qualities and depended on him, even though nothing but contempt was shown. When Colonel George Croghan was called before a Court of Inquiry, it was for personal reasons, inadequate pay and unable to meet his obligations. Whiskey was always available and when pressures were closing in on George, he was too weak to abstain when in the company of drinking buddies.

Colonel Croghan's official reports made between 1826 and 1845 were brilliant. He reported problems found in the remote forts on the frontier and offered solutions. For years, his reports were used by the military powers in D.C. to set up their programs. Some careers were made as a result of using these reports as their own.

The following is from *Army Life On The Western Frontier*, by Francis Paul Prucha, who edited pages and pages of *Official Reports* made by Colonel George Croghan between 1826 and 1845.

> The Inspector general was the eyes and ears of the General-in-chief. There is no doubt that a good deal of the efficient operation and improvement in the army depended upon their careful observations

and frank comments. In Croghan the army had an accomplished soldier and a man whose love of things military made him an admirable watchdog for correct instruction and military discipline. Recommendations which he made found their way into War Department orders and into the new editions of the General Regulations. His long term of service indicates that he was respected despite his aberrations, yet he was not able to exert the influence he desired. There were many forces at work, political as well as military, upon the organization and life of the army over which Croghan had no control.[26]

Jesup made use of all of George Croghan's reports, as well as picking George's mind for information about the conditions on the frontier and asked for George's recommendations. Later, Jesup hired his sons-in-law, Sitgreaves and Nicholson for positions on his staff. He led the two to believe he was the originator of these army reports made at the times he had visited some of the forts. Very soon after Jesup's death in 1860, son-in-law Sitgreaves authored a book giving Jesup all the credit for these reports.[27]

Colonel George was in Philadelphia, Pennsylvania in February, 1844 after moving his family to Philadelphia. He had just heard the news that the House of Representatives was going to reduce the numbers and pay of the army and navy. He was very concerned and sent a letter to William A. Gordon, Esq., Office of Quarter Master General, Washington D.C. The last paragraph in the letter reads, "I left on your desk as you will recollect in April last certain books Gen. Regulations Ord. Ref. etc., which must beg of you to send to me by Harndens Express bus. They may be tied up in a bundle and addressed to me 29 Grand St., Philadelphia G.Croghan."[28]

Colonel George received no favors or benefit for these reports, but others in the Quarter Master General's office obviously would benefit. George would receive nothing but less pay.

As a soldier of the U.S. Army, he was never recalled for dishonoring the Flag of Truce. His brother-in-law, Thomas Jesup was recalled to Washington for failing to honor the Flag of Truce, during Jackson's removal of the Indians from the everglades in Florida and Georgia. After months and months of failing to capture the Indian Ocelo, Jesup tricked him into meeting under the pretense of making peace. When the Indians were seated on the ground, Jesup's army came out of the woods, circled the Indians and captured them.[29]

There is cause to believe that if Colonel George with his military intelligence and leadership qualities had been in charge, he would have made a victorious capture in a short time. Jesup was recalled to Washington for a court hearing. There was no court hearing, but he never regained the respect he had lost.

The St. Louis, Missouri Fur Company and John Jacob Astor had failed to establish permanent posts in the West. Colonel George Croghan succeeded in making the western frontier safe from attacks for twenty years, but others received the credit.

Official Inspection Reports Made by
COLONEL GEORGE CROGHAN
Between 1826–1845

War Department Records of the National Archives

One of the Inspector General's chief responsibilities was to report to the General-in-chief on the strategic location of the army troops assigned to the west. How best to distribute the soldiers was a problem of special concern in an army of small size and large duties. Croghan obviously delighted in this part of his task. He had had considerable military experience and did not lack self-assurance when it came to setting forth his own ideas of what was best for the defense of the frontier. Not only did he repeatedly criticize the locations and the fortifications of the posts, but he was constantly on the alert to catch anything which might be considered a deviation from the strict military spirit which his training and temperament had made a part of him.

In 1833, Croghan added to his regular inspection report a general survey of the military frontier in the West, from Sault Ste Marie to New Orleans, in which he gave an admirable picture of the state of the western frontier on a post-by-post basis. This long report and the shorter selections of this first chapter set the stage for the more particularized items which caught the attention of the Inspector General and which he in turn passed on to the officials of the army in Washington. By Francis Paul Prucha.[30]

Inspector Colonel George Croghan

St. Louis, October, 1826:

"Our frontier posts ought to be viewed as if placed directly upon the lines of a hostile territory and should therefore be prepared for immediate hostilities. The posts should be strong in themselves, the garrison sufficient, well supplied, and throughout that vigilant police observed which would presuppose a state of war. This is far from being the case; it would seem that the purpose was not to operate upon the fears of the Indians by an array of military strength and an appearance of constant watchfulness, but to gain them over by the softer arguments of unreserved intercourse and unsuspecting confidence. I now repeat that which I before assert-

171

ed, 'Our military have lost character among the Indians,' and that it can not be recovered under a continuance of the present system of external police.

Ask any officer at one of those posts what his place is in the event of alarms, and his answer will be, I don't know, no particular one has been assigned to me; we never have alarms, either false or real. Direct the officer in command to receive an enemy that will attack him in a few minutes, and it will be found that he requires half a day of preparation. He has to designate the different stations, to appoint to those particular commands and after all set an enquiry on foot as to the best men to place at the guns before he can discharge a single piece, Order a shell to be thrown, and the time for firing three or more will be taken up in finding one small enough to enter the muzzle of the howitzer, as was the case at Fort Snelling."

"Look at Fort Atkinson and you will see barn yards that would not disgrace a Pennsylvania farmer, herds of cattle that would do credit to a Potomac grazier, yet where is the gain in this, either to the soldier or to the government? Ask the individual who boastingly shows you all this, why such a provision of hay and corn. His answer will be, to feed the cattle. But why so many cattle? Why—to eat the hay and corn—should any of the posts be threatened, where is the disposable force to send to its relief.

I do not say that a soldier shall never be called upon to do duties, unless such as may advance him as a tactician—far from it. I wish him to be occupied and desire only that such service as he may be called to perform, not purely military, may be considered as secondary. I would have the soldier point to his garden in proof of the good provision he has made during the short intervals from military exercise, rather than boastingly talk of his proficiency as a farmer."

George's time on the frontier wasn't an easy comfortable life. While at St. Louis near the close of his first tour, Croghan wrote to his brother-in-law General Jesup:

"I reached this place from St. Peters about the first of this month and would at this time be at or near the [Council] Bluffs but that

I was taken sick of fever which confined me for a week to my bed and a longer time to the house. I have pretty nearly recovered my strength and having a good appetite will be my self again before I reach the confines of the state. I can not say that my tour thus far has been a pleasant one. I have been much exposed to the influences of the weather and almost through-out without a traveling companion. I have been interested, only though in as much as I passed through a wilderness country that I had never before seen. I have visited the posts on the lakes and the Mississippi and in the time have gained enough to satisfy me that the present disposition of the two Regiments which garrison them is as bad and ill judged as it could be—I am wrong, it will be worse when the post at Prairie du Chien is broken up and its garrison removed to St. Peters. Who in name of all that is military, could have advised General Brown to direct the abandonment of Prairie du Chien, the only important military position on the upper Mississippi?"

Washington, December 9, 1833

"Having now completed my remarks upon the state and condition of the posts visited by me since August last, ... defenses when I last inspected them were in a most wretched condition, not a single post along the whole line, with the exception of Fort Independence perhaps, being able to protect itself against the insults of even an armed brig; if any of them had guns mounted, they were upon decayed carriages that could no longer traverse and upon platforms that had long before been declared unsafe.

I confess that the necessity for the occupancy of a point on the St. Marys has not as yet been made apparent to me, and why a garrison should be established at this particular place I am at a loss to conjecture, for in my judgment no substantial reasons can be assigned in support of it's ability. This is no thoroughfare for Indians, nor has it ever been. The country round about offers no inducements to them; it furnishes but little game and is entirely unfit for tillage. Remove the Indian agency (which I would not advise) and but few Indians would be seen on the banks of the St. Marys, unless during the fishing season or when on their annual route to ... to receive their customary presents from the British. If to keep in check the turbulent spirit of the Indians be the object of

distributing troops along our frontier, that would be more effec-
tually served by even a single company at the portage of the Fox
and Quiescence rivers than by an entire regiment here.

It may be asked, shall the post be abandoned. I answer. A mistake has
been committed; to endeavor to correct it by withdrawing the troops
and abandoning the post would create quite a sensation in the village
which has grown up since its establishment, for there would in that
event be in prospect the certainy of an occasional outrage from the
passing Indians."

Administration and Services
Regularity of payments made to the troops was stressed

Fort Towson, August, 1827
"Arms in the hands of the men in the highest possible degree of polish,
but throughout unserviceable, they have here, as at Cantonment Jesup,
a fashion of staining the stock of the musket with vermillion and also
of softening the pan steel, that it may receive the better polish. Both
these practices will be put a stop to."

Jefferson Barracks, August 6, 1831
"If the present mode of cleaning them be persisted in, they will not long
remain."

As a result of George's reports with instructions as how to care for arms
and equipment, his complaints were eventually heeded years later.
Regulations were issued from Washington in the General Regulations
issued in 1835.

"The arms will not be taken to pieces without express permission from
the Captain or other commissioned officer. The practice of highly pol-
ishing the barrels of the muskets will be discontinued; all that need be
required is, that they be kept clean and free from rust, except the bay-
onet and bands, which are to be kept bright. Cartridge-boxes will be
polished with blacking instead of vanish, as the latter cannot be used
without injury to the leather. White lead is forbidden to be used in
cleaning the belts and gloves; it being found to possess qualtities injuri-
ous to health, when near the person."

Hospitals and Medical Care

Fort Howard, June, 1828:

"In looking through the several rooms, my attention was called to the medical library, which as to number of volumes appears well enough but furnishes very little variety. The catalogue stands thus: Bell on venereal 7 copies, Cooper's Surgery 3 ditto, Dispensatory 7 ditto, Dorsey's . . . "

Baton Rouge, May 8, 1844:

"Surgeon [Benjamin F.] Harney seems to pay every attention to the sick, of whom he has at present but 4, none of them seriously indisposed. The hospital appears to be an uncomfortable building, rendered so by its expose situation, being completely uncovered, without trees or shade of any kind, to the scorching rays of the sun. It is true that a gallery extends entirely around the building, but it does not prevent the sun from shining into the sick wards. I have advised Dr. Harney to endeavor to get an awning for the entire west front. The supply of medicines and medical stores is abundant and of good quality."

Food Preparation

"The soldiers at the western posts ate well and the prepared meals had to meet Colonel Croghan's approval. Fresh meat had to be cooked to meet high standards using salt and vinegar—sometimes roasted or baked, but never fried. fresh vegetables of all kinds were supplied in abundance. Food was served not only in as it may relate to quality and variety of the dishes served up but also as to the style of the cookery and the order and neatness in the presentation."

Clothing

His heart went out to the men in the West. Heavy fatigue duty in the building of roads and the clearing of land took heavy toll of their regular issue and to replace their clothing was to purchase at their own expense. He did not overlook the importance of a proper fit. A parade could be spoiled poorly designed and ill-fitting uniforms.

Nothing pleased the old soldier Croghan more than a fine appearance of the soldiers on parade, and a special section of each report was reserved for his comments on "Appearance under Arms." He found a great variety. Some commands appearance he delighted to see—good-looking specimens, uniform in size, and quick and graceful in their movements. Others were awkward, unmilitary. In one command, 32 out of 52 were immigrants who could hardly speak enough English to be understood. Croghan was disappointed not so much as a censure, the ones the army had to rely on for its regiments."

Fort Leavenworth, August 15, 1831:
"The pantaloons issued to the soldiers are too small and short."

Fort Crawford, August 10, 1838:
"I would again express my regret that a change was made in the uniform of 1812."

Fort Crawford, July 11, 1842:
"The long tail of the present coat is a useless appendage and a ridiculous looking one."

Fort Leavenworth, August 16, 1842:
"The Infantry Company appeared on review in full uniform, the abominable long tail coat with wings, but afterwards in the round blue jacket and white pantaloons."

Fort Leavenworth, August, 1843:
"It is wished that the white cotton pantaloons should give place to some dark woolen or other stuff. The present white is so easily soiled that in 10 minutes after the men are mounted, it becomes, by the rubbing of the horses against each other, so stained as to become offensive to the sight, besides which it is totally unfit for use upon the long and frequent excursion over this prairie region, where the dews are heavy during the warm months."

Books

Fort Snelling, August, 1828:
I would have it understood that the regimental books and such company books as have been exhibited are to the fullest sense of the term the property of the regiment, not having been furnished by the Quartermaster Department but made at the post out of such paper suited to the purpose as could be purchased of the sutler. . . . each receiver of a set was left to follow his own understanding of the expression of the regulations.

Fort Snelling, August 17, 1834:

None of the books, I am sure, exhibit incorrect entries, but the books of no two companies are kept after the same form.

Fort Crawford, September, 1840:

Books. Correctly kept: and here let me suggest the propriety of having all regiment and company books that may hereafter be issued ruled off and leaded precisely alike if it be considered of consequence that the same forms of entry be adopted throughout.

Temperance Society

Baton Rouge, May 8, 1844:

A post temperance society has been organized, over which one of the non-commissioned officers presides, and already one half of the men have enrolled themselves as members.

CHAPTER SEVEN

Mexican War

D r. John Croghan was surprised when nephew St. George made a decision to return to Locust Grove as a farmer. George and Serena's marriage had ended and perhaps St. George was swayed by his father's efforts to reform and he wished to be near his father offering support and encouragement. St. George at that time had not yet married.

The three brick houses on "Croghans Row," built by Grandfather Croghan, now belonged to St. George Croghan. Each house having three stories and attic were located on the corner of Fifth and Main Streets. Uncle John had expected St. George to return to his family roots in Louisville to practice law. Upon arrival, he immediately started making changes to have the three brick houses demolished and rebuilt, using the same floor plans as before for each house. Beautiful workmanship was applied and nothing but the best material was used.[1] He was planning to rent two and live in the other. Charles Thruston was property agent for the Croghan family's Louisville property including William Croghan and the Jesups. At one point, Thruston reported to Dr. John that St. George was extravagant in building the houses and using too much of his inheritance.[2] It seems the Clarks and Croghans were all so closely connected that one's business became everyone's business and gossip never ceased.

In April, Colonel George was at Mammoth Cave in charge of managing the cave for John, ordering supplies and planning for necessary repairs to buildings. Life seemed to be better for George and he was still attending Temperance Society Meetings.

Suddenly, the news of Ann's death reached Colonel George and immediately he went to be with the Jesup family in Washington. Dr. John's poor health prevented him from traveling. William Croghan was on an ocean steamer returning from Europe.

My dear Genl. . . . This morning I received a letter from Doct.
Miller announcing the death of my beloved sister. I am distressed
beyond measure at this unexpected & melancholy event. My only
sister is gone, gone to a happier & better world. My grief is that
of a sincerely attached brother. Most deeply do I sympathize with
you, dear Genl. And distressed nieces & nephews. At present I can
write no more on a subject so painfully afflicting to me. My love
to my dear nieces & nephews—I am dear Genl.—Most truly
yours—John Croghan.[3]

Ellen Pearce Bodley, descendant of Jonathan Clark wrote the following
from Vicksburg: "Sister received dear Martha's letter yesterday in which she
mentions cousin Anne Jesup's death. I feel it to be a family affliction and
Oh how much Gen. Jesup and her children will feel her loss. I do really feel
for them. She was to me one of the most perfect women I ever knew."[4]

Ellen's husband, William Bodley wrote on the 15th, "Last night I visited Genl. Jessup's and saw him & his daughters & Col. Geo. Croghan."[5]

Immediately after losing his sister, Colonel George was called to active
duty in the Mexican War. Any plans for father and son to spend some
years together would have to wait. The year 1846, was also an eventful
year for weddings in the Croghan family which Colonel George would
miss. In January, Mary Serena Jesup had married James Blair. Colonel
George and Serena's oldest, Angelica, married Christopher Wyatt and son
St. George married Cornelia Adelaide Ridgely in Baltimore, July 1, 1846.
Their youngest, "Tinie," was still with Serena.

St. George's bride was a Livingston relative. Cornelia was the great grand
daughter of Chancellor Robert Livingston. The following chart shows how
confusing these marriages were.[6]

John R. Livingston	— Brothers —	Chancellor Robert Livingston
m.30 May 1789		m.9 Sept 1770
Eliza McEvers		Mary Stevens
Serena Livingston Croghan	— 1st Cousins —	Margaret Maria Livingston
m. May 1817		m.1798
Col. George Croghan		(Cousin) Walter Livingston
		Cornelia Eliza Louisiana Livingston
		m.1823
		Capt. Charles Ridgely
St. George Lewis Livingston Croghan	—	**Cornelia Adelaide Ridgely**
(*B. April 1822 D. Nov. 1861*)	**m. 1 Jul l846**	(*B. 12 Feb 1827 D. Oct 1857*)

After St. George's wedding, he and Cornelia lived in one of the three brick houses. Dr. John was in very poor health, unable to run the farm and nephew St. George had taken on the responsibility for his uncle before his wedding. After several months, John was of the opinion that St. George had failed miserably as a farmer at Locust Grove. He wrote to family members that fences needed mending and St. George had always lived in the city and he had made trips to New York at the time spring planting should have been taking place at Locust Grove. The doctor had suggested he should return to Philadelphia.[7]

Dr. John was responsible for allowing the farm to become run down by sending the slaves to work at Mammoth Cave.

In the meantime, Col. George wrote to General Jesup, that Monterey was three days of fighting, losing a lot of brave and gallant men. He hoped General Taylor would not advance beyond Saltillo. There was little likelihood of his being placed in position to authorize such a movement.

October 1847, George was still in Monterey and since May, either the ague, fever or the diarrhea had a hold on him. He received a letter from Doctor John, who reported a difference of opinion with Serena as to who owes whom in regards to financial transactions. George was distressed and astonished at what had taken place and went on to say that Serena certainly had forgotten that after a careful examination with him over John's accounts in 1843, records showed John was the creditor and not the debtor. When George arrived in Mexico, he weighed 168, but had dropped to 148 lbs. As soon as the City of Mexico should be taken, General Taylor had told him he might leave the country. He anxiously waited for the War Department's permission to return to the United States.[8]

George was given a sixty day leave from the Mexican War, but it was a long slow uncomfortable journey traveling with supply wagons especially since he had suffered for many months with intermittent fever. It seems once he reached Locust Grove on 15th December 1847, he had never been so happy to arrive at his old home. He found the Doctor was not well, but reported that St. George, wife and little daughter were fine and was happy to know that St. George had been very industrious and quite the farmer.[9]

Tinie had traveled from Philadelphia to meet her father Colonel George in Washington. St. George and his family were there for a family gathering, but the visit was interrupted when suddenly Colonel George was ordered back to Mexico.

At the end of March, Tinie had returned to her mother in Philadelphia and St. George returned with family to Louisville after deciding against joining the army. Doctor Croghan was encouraged and believed he would then be an industrious farmer.

Colonel George was in Mississippi with General Zachary Taylor, his wife and daughter in September. Laying around and doing nothing made Colonel George very impatient and he was wondering what he was doing there. Furthermore, he was surprised to find that General Taylor had not mentioned his name in any of the writeups. He had fought in the battles and felt he had shown bravery as much as any of them, but no deference was shown him.[10]

He had joined the army on its march to Monterey, and was present at the assault of the place. During the crisis of one of the three days' fighting, a Tennessee regiment shook under a tremendous concentric fire, Croghan rushed to the front and, taking off his hat, the wind tossing his gray hair, shouted:

> Men of Tennessee, your fathers conquered with Jackson at New
> Orleans—follow me! The stirring words were received with bursts
> of cheers, and the troops, reanimated, dashed on to victory.[11]

John received a letter from another officer in which he wrote about Colonel George taking a brave lead in another battle when the Kentuckians fell back and there was a cry of " 'Croghan—Croghan,' and on they went at once to victory—his name was enough."[12]

By September, 1848, Colonel George's patience had reached its limit after the army had waited and waited for supplies and transportation, but yet 8000 volunteers had arrived with nothing to do. He sent a letter to brother-in-law Thomas Jesup expressing his frustration and not necessarily blaming him, but keeping him appraised of the inexcusable situation.

> Pascagola (Miss.) 1st Septr. 1848
> to Majr. Genl. Qr. M. Genl. Washington, D.C.
>
> Dear General
>
> I came hither from New Orleans with Genl. Taylor, his wife,
> daughter, a D.C. Bliss &c. on the 17th. Augst. He talks of return-
> ing to Baton Rouge on the 15th of this month, but how long I am
> to remain I know not, as that must depend on the wise ones in

Washington who act strangely sometime. I was hurried off from that place by an order from the Secy of war [Jesup's friend] requiring to report in person to Genl. Taylor, & to this hour, after waiting anxiously for the last 8 weeks I am ignorant of the purpose of the order, nor can Genl. Taylor himself conjecture the object of the Secy. in sending me hither.

Never have wise men before, acted so unwisely as these did in establishing a camp on this coast. I'll take care in one season to prove the fact & at the same time to show up other acts of like wisdom & foresight. My love to each one of the family

Your friend G.Croghan.[13]

That letter caused Jesup to explode. Francis Paul Prucha summed it up very well. "George had mustered in troops and took an active part in the war himself with the army under General Zachary Taylor. It was not a happy time. He was ill, dropping from 168 to 148 pounds in weight in the two weeks after arriving at Monterrey, and he was worried by the financial problems in the family estate. He expressed great eagerness to leave Mexico."[14]

The 8,000 men had arrived and everything was paralyzed for lack of transportation—not enough of anything, especially steamboats. Somewhere along the way there had been no planning or preparation.

George's letter wasn't the only one Jesup received about the waste of time, money and soldiering. "Taylor's impatience, especially against the office of his old friend, quartermaster General Thomas S. Jesup, was to grow into a festering grievance. Throughout the war was the inescapable time lag in the transmission of reports, dispatches and information, and the return of reports of action taken."[15]

Steamer Genl. Worth near
Baton Rough 4th Novr. 1848

Dear General

In the name of all that is wonderful, what is there in my communication to you of the 1st that it should be adjudged indecorous? I am astounded, and deeply mortified at finding that you know so little of me, as to imagine, even for a moment that I could say or even think anything, in which your feelings might be interested, wanting in the profoundest respect and regard. Look again into the case and you will I am sure see it in a different light, and a favorable one.

You sent me a paper written by a man in Philadelphia to the President, complaining that he had written many letters to me & that no answers had returned. I had just cause against the fool. I will be in Louisville about the 20th. Inst. On a short visit to the Dr.

G. Croghan.[16]

George didn't make that return home. Instead, it was sad news that reached many old friends and loved ones in Kentucky.

On January 8, 1849

We copy from the N. Orleans *Picayune,* of the 9th, the following notice of the death of Col. Croghan. Dr. Croghan, a brother of the Colonel, died at his residence in this county, of consumption in less than three days after Col. C. died at New Orleans of cholera. The Dr. was a man of considerable property, and the most of it we understand was left in trust to the Colonel:

"Col. George Croghan,—This distinguished officer died last evening from the effect of a disease resembling cholera, which he suffered to remain upon him for near two days before calling in medical assistance. He was attacked on Saturday, but paid no heed to his complaint till toward noon on Monday; when his symptoms became so distressing that medical aid was summoned, but no relief could be had from the most skilful and considerate treatment. He lingered till night-fall, in the perfect possession of his faculties When he felt the hand of death upon him, he gave directions as to the disposition of his body, with the greatest calmness, entrusted various messages with surrounding friends, and closed his eyes forever.

Upon the breaking out of the last war, Col. Croghan entered the army. At the early age of nineteen he made the gallant defense of Fort Sandusky. By this brilliant feat he inscribed his name upon the scroll of fame. He married and resigned his commission shortly after the peace. But during the administration of Gen. Jackson he returned to the service with the commission of Inspector General, which was tendered to him by that illustrious commander. He held this office up to the time of his death. He was in his fifty-ninth year and leaves behind him a wife and family.

It was scarcely hoped that he would live through the day yesterday. It was the glorious Eighth of January, and as the booming of cannon would shake the chamber of death, thoughts of the olden time would come over him, and he would straight-way revive. He had heard such sounds long ago, and they spoke to him of the past. Towards evening he weakened as the moments wasted. He struggled through till night closed upon the earth. The military had fired their last salute in honor of the expiring day.—When its echoes had ceased to reverberate, the hero of Sandusky was dust."[17]

Three days later, January 11th, Dr. John Croghan died.

Louisville Jany. 18, 1849

My dear Sir [General Jesup]

You will have been informed by Telegraphic dispatch and from other sources of the death of Doct Croghan which occurred on the Morning of the 11th inst. A messenger reached me at four oclock on Monday [illegible]—from the Doctor, requesting me to go out, I got to him at dusk and found him as I thought dying, he ... lived until 35 minutes after six on Thursday morning. I left him at half past four believing him to be decidedly better. Although as himself, he certainly [illegible] slept better than he had for some nights previous. I had [illegible] gotten to sleep however before Judge Brown came up into my room to announce his death. The Judge informed me that after I left the Doctor he rested very well until half after six, he then [illegible] the judge taking a bead in the room and as he was getting to bed the Doctr. called his Boy, Isaac who was asleep in a chair at the fireplace. The judge got up and sent the Boy to his Master—just as he got to the bed side, the Doct. was seen to throw up his hands and before the judge got to him he was no more. Every proper attention was [illegible] to his remains and on Friday the 12th he was buried in the family burying ground.

As Col. Croghan was expected from the South any hour when his Brother died, I had the Home secured and every thing in it, taking away only some papers, and left it in charge of the gardener in whom the Doct. had every confidence, intending to do but nothing, until the Colonel should arrive. We have [illegible] received this morning the melancholy intelligence of *his* death also. This even will require an [illegible] action on the part of the Trustees....

The will was formed from other wills & parts of wills & under the Doctors direction its execution was but a few hours prior to his death. ... Geo. C. Gwathmey.[18]

In the Mexican War, dysentery struck down men by the hundreds. The dead march "was ever wailing in our ears," one writer noted, while another declared that it became a sound so familiar that even the birds learned it. Funeral escorts "to that vast and common cemetery, the Chaparral, and the crashing of the three volleys fired over the grave of some soldier, shroudless, coffinless, only in a blanket, almost ceased to attract attention."[19]

General Zachary Taylor was sworn in as President on 5th of March, just two months after George and John Croghan's death. J.F. Taylor, (the President's brother), answered a letter just received from Major John O'Fallon, son of Frances Clark, "Well, I should like to see you once more and talk over our boyish days as there are but few of us left now of the many boys who were our playmates in our young days, for death has thinned them off to a fearful extent. It is melancholy to think of the large family left by your Uncle, Major Croghan, and but one is left of them. J.F. Taylor"[20]

One year later, 1850, the last Croghan (William) died. Zachary Taylor died the following month. Taylor's life as President was cut short.

After Dr. John's death in January 1849, St. George inherited Locust Grove. The household furnishings were to be divided between Ann Jesup's children and Colonel George Croghan's children. St. George paid the Jesups for their half interest in the furnishings. Most of the items such as family treasures and items with historic value had already passed onto Ann Jesup and William Croghan Jr. However, the Locust Grove Estate Papers list included fine pieces of furniture.[21]

Serena and Tinie moved from Philadelphia after Colonel George died and perhaps joined St. George and family at Locust Grove, especially in July when Colonel George's funeral and burial took place in Louisville at Christ Episcopal Church, July 10, 1849. William Croghan, Jr. must have been present when brother George was buried in the old Locust Grove family Cemetery.

CHAPTER EIGHT

Serena and Tinie on Their Own

In the fall of 1850, Serena and Tinie were spending the winter with the Jesups in Washington. Mary Blair received a letter from her father in which he goes into great lengths explaining the situation at home. He is careful to say he doesn't spend much time at home—leaves early in the morning before others are up—has his lunch sent to his office—comes home after everyone has retired for the night. He goes on to say, "Mrs. Croghan and Tiny will be good for the girls this winter and all are writing letters to you."[1]

A few months later, Serena's father John Livingston died September 25, 1851.[2] The youngest sister Margaret Livingston Lowndes inherited Edgewater on the Hudson River. Serena had been given the Louisiana plantation when she married George, but it was lost to creditors. Many years later, grandson George Croghan III wrote that Serena had money, but her trust in others deprived her of most all of her estate.[3] Tinie Croghan turned eighteen in 1852. It is possible she received some inheritance from her Grandmother Eliza McEvers Livingston.[4]

Serena and Tinie sailed from New York in 1853 around the Cape for San Francisco to join Angelica Croghan Wyatt and her young family. Only the wealthy could afford to go the route around Cape Horn. Others went across Isthmus of Panama which was a very uncomfortable and difficult way to travel. The clipper ships had their heyday from 1850 to 1854. This era started in 1848 when gold was discovered at Sutter's Mill, California. There was a wild rush west.

> There were dangers of sailing around Cape Horn which was the southern extremity of South America. It had an Antarctic climate and frequent storms. The ship would be on the water for six months in 1850. The travelers were often nauseated and bored stiff. Passengers

had to survive tropical heat at the equator, icy gales at the Horn. Sometimes, there was play acting, debates, musicales, readings and Sunday preachings.[5]

Angelica and husband, Christopher Wyatt, Rector of The Episcopal Church of San Francisco, must have been relieved with tears of joy when they arrived safely.

If Serena and daughter expected to see Mary Jesup Blair when they arrived, they were disappointed. Mary had sailed for home only a few weeks before. After Mary and the two children had returned to Washington City, James Blair's letters revealed how lonely and sad he had been and the arrival of Serena and Tinie from back home was comforting.

The presence of Serena, Tinie, Wyatts and other family friends from back East seemed to have brought James back to life. He very soon, began to escort Serena and Tinie to church and social events. Serena stayed busy with sewing and must have been skilled in fine needlework. Serena's great-great granddaughter Virginia Wheaton had a silver pitcher, approximately 12 inches tall, which was awarded Serena with the inscription:

Awarded to Mrs. S.L. Croghan
For Beautiful Embroidery
Warren & Sons Exhibition—San Francisco
October 1853

There were other transplanted families who had been part of their social circle back in D.C. and were now, residents of San Francisco. James Blair's letters included news about these families, the Wyatts, Serena and Tinie. Excerpts from the following letters give a picture of that bustling time in San Francisco.

Getting married and having parties appears now to be the rage in California. I have accompanied Mrs. Croghan & Tiny out to three parties of late. The weddings produce these latter festives. I went to one last night at a Mrs. Beals on the Hill, corner of Stockton and California. Mes. Croghan was very much alarmed in the carriage,— screamed and that startled Tiny, who answered in finer or higher, but when I told them 'that it was in the sand and it wd not hurt and wd even rub off before it got dry' they laughed heartily in the midst of their fears, especially as it was only the plunging of the horses dragging a heavy carriage up the hill in sand.[6]

James frequently spent an evening at Mr. Wyatts and escorted Serena and Tinie to social events. He wrote that getting married was all the rage in town and is afraid Miss "Tinie" Croghan will give way and take the fever as she was so much surrounded with beaux. He amused himself sometimes by watching the progress of events to see who gets the advantage.[7]

Augustus Rodgers was one of those who had romantic intentions for Lutie (Lucy Ann) Jesup, but Gus noticed Tinie's personality, beauty and fell in love with the charming girl.

December 15, 1853, James Blair suffered a ruptured aorta and died. His brother, Montgomery Blair, sailed to California and brought his body home. From all indications, Serena and Tinie returned with Montgomery on the trip back to Washington D.C. with James' body. The Wyatts could have returned on the same boat with Montgomery Blair. It was late in 1854 when they arrived back East.

Three years after returning to the East from California, Serena and Tinie were faced with more sadness. St. George's wife, Cornelia Adelaide Ridgely Croghan, died October 9, 1857 in New York. The four motherless children ranged in age from three to ten.

The following summer, 1858, twenty-four year old Serena Croghan (Tinie), married Augustus F. Rodgers, a Washingtonian. They were married in St. Johns Church, Lafayette Square, Washington D.C. Very soon, Augustus had to return to California, leaving Tinie at Serena Croghan's place, near Kingston on the Hudson River. He had been sent out to California by the Coast and Geodetic Survey to map the unrecorded inland country. Baby Cornelia Rodgers was born in Ulster County New York.

Serena wrote to Augustus' mother, Mrs. Commodore Rodgers, Washington D.C., in which she tells how she had put her "darling Tinie" with her baby, "Daisy," on the boat for a long journey to California, alone.[8] Serena stayed in New York with St. George where she was needed.

The following is taken from an eighteen page letter received by this writer from Virginia Wheaton, May 1990:

> Very courageously, Serena ["Nana"] and her baby set forth alone on the long long trip to California. . . . Nana thrilled me from the time I was a little girl with her stories of crossing the Isthmus of Panama on mule back while she carried her baby in one arm and held an umbrella in the other hand—the mosquitoes were awful. They spent the night in a room—up over a bar room in a simple

wooden Shack where there was carousing all night. She certainly must have had some of the pioneer spirit in her veins because in all her stories, there was never a complaint, a moan or a groan. It was always just a Tale of adventure & excitement and at times discomfort.[9]

During the years 1960–1964, Ms. Emilie Smith was in close contact with Virginia Wheaton who lived in California. Upon hearing the news that the Croghan tea service had been found in England and would be returned to Locust Grove, Virginia wrote,

> The news of the Schenley descendant in England interested me very much—think it was one of her family, Baroness Hermione von Solvens, who sent my Grandmother (whom she had never seen) a check for $50,000 to buy a house in S.F. [San Francisco]. This was when she heard my Grandmother had been widowed at 22, before Mother was born, and that my Grandmother had taken a job at the mint—an unheard of scandal for a *Lady* and widowed, considered a disgrace to the whole family. My Grandmother gave the house to her Mother and Father, Capt. and Mrs. Rodgers (she was a Croghan) and my grandchildren were the 6th generation to stay in the old place which we finally sold three years ago. My memory is that Baroness von S. was a daughter of my Great Grandmother's cousin, who was the girl that ran off with Schenley.[10]

Virginia Wheaton also wrote that "Nana" was a most fascinating gentle and lovely person and always serene. She was still amazed at how tiny Serena Rodgers was never weighing more than 105 or 106 lbs. Embroidery and reading was her past time. There was no automobile and the family house was on one of San Francisco's steepest hills. She had to walk up or down for two long steep blocks, in order to get the little cable car, which didn't seem to faze her. In rain or shine, she never missed a Sunday at church.[11]

Serena Croghan Rodgers lived until 1926, but after her husband's death in 1908, she never wore anything but mourning with white cap in the house and black cap and veil for outside. She sent to London to the same shop where Queen Victoria had her caps made. Virginia distinctly remembered a round polished stick which is associated with these in her

Four generations of descendants of George and Serena Croghan. Center, their youngest daughter, Serena (Tinie) Croghan Rodgers; right, granddaughter, Cornelia Livingston Rodgers Nokes; back, great-granddaughter, Virginia Rodgers Nokes Murphy; front, great-great-granddaughter, Virginia R. Murphy Wheaton. Photo taken in 1913 or 1914, in Ft. Mason, California (courtesy of John Wheaton)

mind. Although Virginia never saw it done, she imagined the caps had to be laundered and starched. The veil folded into pleats and the cap may have had a little fold (as that in "slouching") at the front and sides into which the rod was inserted for drying.[12]

Virginia's grandmother, Serena Rodgers, spoke often about her father's young days, his defense of Fort Stephenson and also, about George Rogers Clark, William Clark and the Indians. Many questions were never asked about Serena Rodgers' mother, Serena Livingston, 1794–1884, "who divided her time between our house and that of her granddaughter Lucy Croghan Brown."[13]

CHAPTER NINE

Civil War—Orphans—California

War was on the horizon in early 1861. Major Robert Anderson, a cousin of the Clark and Croghans, was the officer in charge when the first roar of guns started at Fort Sumter, South Carolina, April 13, 1861. Again, Serena had no way of knowing what was in store for her and what an impact this event in South Carolina would have on her life. The following account tells the story of the tragedy that struck.

> On the evening of November 11 the enemy made strong demonstrations . . . an attack on the next day, General Floyd ordered the army to fall back three miles. Next morning enemy advancing to Fayetteville . . . Gen. Floyd ordered a retreat. On the morning of 14th . . . the enemy were near and rushing on the brigade. At this the cavalry, under command of Colonel Croghan, were ordered back to scout the country and ascertain the enemy's distance. When they had gone back two miles they met the enemy's distance. When Colonel Croghan ordered his men all to dismount, though he did not, when the pickets of the enemy fired on him, and he fell mortally wounded. His men took him up and carried him some two hundred yards to a house, when they discovered that the enemy were closing in, and the colonel told them to fly and save themselves, for he was dying.

> At the moment those who were with the colonel discovered that their horses had been taken by the Yankee pickets, who had rushed upon them, they turned and fled, and the whole cavalry came within five minutes of being all cut off and escaped. The calvary then all swept on in abreast until they came up with the fear of our infantry, and proclaimed that the enemy were pursuing in double-quick time. Then appeared a scene in our army indescrible, and of terrific confusion. At the word "the enemy are pursuing" all broke

off in a wild run, some so frightened that they threw away their knapsacks and all they had, but gun and knife to defend themselves with. It required great effort upon the part of the officers, who were somewhat cool, to prevent a perfect rout. After this day the brigade continued its retreat, but with a great deal of toil and difficulty, and finally encamped here on the 24th of November. This encampment is near Peterstown, in the south edge of Monroe County, and it is expected that the brigade will winter near her.

Colonel Croghan fell into our hands mortally wounded, and died in a few hours. His body was sent by General Benham to the confederate commander, with a note hoping that he would appreciate the desire thus expressed of mitigating the horrors of war. He was Kentuckian, the son of the George Croghan who, in 1813, with only 160 men, defended Fort Stephenson in Ohio against 1000 British regulars and Indians, and who, a quarter of a century later, received the thanks of Congress and a medal for his gallantry on that occasion, and died as inspector general of the United States army.[1]

Colonel Rutherford B. Hayes was not too far away from St. George when he was injured. They both were in the same battle, but fighting on different sides.[2]

The most devastating, bloodiest war in America's history had ended April 1865. Serena would, no doubt, feel her life was in the past and never feel quite whole, again. The Clark cousins also, suffered a great loss in that war. Playmates of long ago, had fought each other on the battlefields. While husbands of the South were fighting to protect their lands, the wives left at home witnessed the horrors of war and never forgave the Union Army from the North.

Serena and the four Croghan orphans must have lived in Washington or New York during the last four years of the Civil War. She had lost her only son, and one of her two daughters lived far away in California. After the war ended in April, 1865, at last, it was safe for Serena to travel with her grandchildren around Cape Horn. It had been seven years since she had seen her most precious Tinie and now would be with her. Serena left the eastern shores never to return.

Three years later, after arriving in California, granddaughter Cornelia married in 1869. Her husband was Englishman Horatio Horner, and a daughter, Mary Sophia Horner was born in 1870. Cornelia's young husband died very soon and she died in 1878, leaving eighty-four year old great grandmother, Serena Livingston Croghan to care for eight year old Sophia.

Serena's four orphaned great grandchildren were very well educated and the second daughter, Lucy Croghan Browne was talented. She was an accomplished pianist and composer of children's songs which were published. Her cousin, Mary Croghan Schenley also, had copies of Lucy Browne's songbooks printed in England. One of those is found in The Filson Club, Louisville, Kentucky.

George Croghan III (1852–1911) was nine years old when his father was killed in the Civil War. He did not gain title to Locust Grove until 1868 and then sold the old family home to James Paul, December 28, 1878 for $28,000.[3]

Serena's last years were spent in a boarding house with her great granddaughter Sophia Horner. However, Serena died at Tinie Croghan Rodgers' home in Oakland. Her grave is located in the Mountain View Cemetery in Oakland, California. It seems proper that she should be buried back in her beloved Hudson River Valley—the sky, mountains and river scenes where she spent her youth with sisters, brothers and cousins in the long cold winters, sleigh-rides, large happy parties and in an elegant new-fashioned vehicle, boats full of merrymakers in summer.

Colonel George Croghan is buried in Fremont, Ohio where stands the tall shaft Monument honoring the heroes of war. While viewing the soldier Colonel George Croghan mounted on top and looking out, thoughts return to "Brave fellows, you have all done a main business."

Colonel George and Serena Livingston Croghan's marriage lasted for twenty eight years. It was said that Serena spoke well of her husband, even though George and Serena didn't have a smooth happy marriage, their love and care for their children was stronger than any human weaknesses.

Hero Honored at Fremont, Ohio

Fremont, Ohio's hero, Major George Croghan had been heralded as the decisive action of the War of 1812 in the Northwest Territory. His memory had always been cherished for his brave defense of Fort Stephenson. For many years, the people of Ohio had no knowledge of where their most famous hero Colonel Croghan was buried. In 1906, Colonel Webb C. Hayes (son of President) made a search in conjunction with the war department to locate and return the famous dust of Colonel George Croghan to the scene of its greatest triumph.

Family members, Christmas 1960. Christopher Ward, Jane Wyatt Ward, Michael Ward, Christopher Ward, Jr. (courtesy of Jane Wyatt Ward). Jane Wyatt, a screen actress and television star, is the great grand daughter of George and Serena Livingston Croghan.

Colonel Croghan was supposed to lie in one of the thirty seven cemeteries in New Orleans where he died of cholera on January 8, 1849, his spirit taking flight just as the last gun of the national salute commemorating the 34th anniversary of Jackson's victory, was fired.

After the quartermaster general gave up on the search as fruitless, a letter was received from a daughter of St. George Croghan, Elizabeth Croghan Kennedy. She had written to her aunt Serena (Tinie) Croghan Rogers, daughter of Col. George and learned from her that he was buried in the beautiful old Croghan estate, Locust Grove on the Ohio River.

From: *Fremont Messenger* August 2, 1906

"Colonel Hayes in company with R.C. Ballard Thruston and S. Ballard at the Kentucky Historical Society, proceeded to the old estate in an automobile, and located the burying plot about 300 yards from the mansion. Thickly overgrown with beautiful myrtle were the moss-covered tombstones of Major William Croghan and wife, the parents of George Croghan. Over in one corner lay an over-turned headstone on which appeared the inscription, Col. G. C. marking the long-sought resting place.

Arrangements were at once made for the disinterment by Messrs. Ballard and Thruston who, with their wives and Miss Mary Clark, of St. Louis, were present, all being related to Colonel Croghan through his mother, of the great Clark family.

The mahogany casket, found at a depth of six feet, was badly decomposed, but the leaden casket within was intact, being six and one-half feet in length, 20 inches wide and eight inches deep. It was immediately boxed and taken to Louisville and thence directly to Fremont, arriving here Sunday evening.

The flag-draped case now lies in state in the city hall. Accompanying the remains are the head and foot stones, pieces of the mahogany casket and a box of the luxuriant myrtle which covered the little burying ground, 48 feet square, to a depth of a foot. This will be placed on the final resting place."

Samuel Thruston Ballard (courtesy of Charles Todhunter)

Rogers Clark Ballard Thruston (courtesy of Charles Todhunter)

From—*Fremont Daily News,* July 16, 1906:

THE CASKET WAS OPENED
And In It Were Found The Bones Of Major Croghan

Sunday afternoon at 4 o'clock the Croghan day executive committee, including Major Tunnington, Col. Hayes, Judge Coonrod and Dr. Stamm opened the shipping case containing the leaden casket which holds the remains of Major George Croghan and in order to ascertain the required dimensions for the stone vault in which the remains are to be interred it was decided to open the leaden casket, which proved to be well preserved. It was wrapped with the mildewed remains of the Stars and Stripes, the soldiers' last blanket. The body had returned to dust, but the bones and underclothing in which it was clad were well preserved by the damp proof casket. It was expected by some to find Major Croghan clad in uniform, but probably owing to the fact that his disease was due to cholera the last funeral rites were of such a nature as to preclude much preparation. The remains will not be removed from the leaden casket, but will be re-interred in a stone vault covered with massive granite slab.

The people of Sandusky County erected this monument to all soldiers falling in the various battles for freedom and for Democratic ideas and ideals. If one pauses to examine the lovely shaft topped by a soldier looking north toward the approach of the British during the War of 1812, the inscriptions give the story and the reason for its existence:

North side:

To Him Who Hath
Borne the Battle
and to His Widow and His Orphans
Erected by the People of
Sandusky County 1885.

East side:

Liberty and Union Now and Forever,
One and Inseparable
1861–1865

South side:

In memory of the
Victorious Defense of Fort Stephenson
on this spot
By Major George Croghan and the
Brave Men of His Command
August 2, 1813.

West side:

Representation of a G.A.R. Badge
Inscription:
Vacant places at our camp fires,
Mutely tell of comrades dead,
Fallen in the line of duty,
Where the needs of battle led.

Memorial to Col. George Croghan in Fremont, Ohio (courtesy of the Ohio Historical Society)

The Bivouac of the Dead

The muffled drum's and roll has beat The soldier's last tattoo!
No more on life's parade shall meet The brave and fallen few.

The Fame's eternal camping ground Their silent tents are spread,
And glory guards with solemn round The bivouac of the dead.

The neighing troop, the flashing blade, The bugle's stirring blast,
The charge,—the dreadful cannonade, The din and shout, are passed;

Nor war's wild notes, nor glory's peal, Shall thrill with fierce delight
Those breasts that nevermore shall feel The rapture of the fight.

Like the fierce Northern hurricane That sweeps the great plateau,
Flushed with the triumph yet to gain, Come down the serried foe,

Who heard the thunder of the fray Break o'er the field beneath,
Knew the watchword of the day Was "Victory or death!"

Rest on, embalmed and sainted dead, Dear is the blood you gave—
No impious footstep here shall tread The herbage of your grave.

Nor shall your glory be forgot While Fame her record keeps,
Or honor points the hallowed spot Where valor proudly sleeps.

Yon marble minstrel's voiceless stone In deathless song shall tell,
When many a vanquished year hath flown, The story how you fell.

Nor wreck nor change, nor winter's blight, Nor time's remorseless doom,
Can dim one ray of holy light That gilds your glorious tomb.

Theodore O'Hara

PART III

Mary Croghan and the Englishman

CHAPTER ONE

O'Haras of Pittsburgh

On January 28, 1823, William Croghan, Jr. and Mary Carson O'Hara's marriage was performed by Reverend Francis Herron in the First Presbyterian Church of Pittsburgh.

Mary Carson O'Hara was the second bride to marry one of the Locust Grove Croghan sons. Serena Livingston had arrived five years earlier. These two wealthy and high society girls married two of the handsomest eligible young bachelors of early nineteenth century society.

Mary and Serena had lived in surroundings of grandeur that only wealth could provide. Much of their time was spent in drawing rooms and attending elegant grand balls. Both young ladies had been referred to as "belles of society." It was quite a transition when leaving that carefree life in the big Eastern cities and being thrust into a role of wife and mother so soon after the wedding. Women's diaries tell a story depicting no smooth transition from the role of belles to motherhood. This change, most often, involved nursing babies through dreaded sickness which ended in death. Too many small graves appeared on the scene in those days.

Even though seasoned travelers were impressed with the elegance and fine living in Kentucky, opportunities for cultural enrichment on the Kentucky frontier were not the same as the larger Eastern cities had to offer.

Mary's father, General James O'Hara (1752–1819) of Pittsburgh, was a native of Ireland, emigrating to Fort Pitt in 1773. He was an Indian trader before entering the army in the Revolutionary War, serving as Assistant Quartermaster, 9th Virginia Regiment. After the war, he purchased land all around Fort Pitt so that no matter which way the city grew, he would have land in the way. Even in a major depression, he managed to hold onto this land, believing that some day his heirs would benefit.

General James O'Hara led the way in business ventures, which included transporting salt from New York to Pittsburgh; built the first O'Hara

Glassworks west of the Allegheny Mountains; established the first brewery; engagedfur trading; shipbuilding, and the operation of a sawmill, tannery, grist mill and dry goods store. As a prominent citizen, he served as a trustee on the boards of banks, church and the academy.

In 1783, O'Hara married Mary Carson (c1761–1834), known as "Pretty Polly", the daughter of a well known Philadelphia innkeeper. Their fortunes grew and they lived on the most fashionable street. Their six children were William Carson O'Hara, James O'Hara, Charles O'Hara, Richard Butler O'Hara, Elizabeth Febiger O'Hara, Denny and Mary Carson O'Hara (c1803–1827). Only three children outlived their parents.

A glimpse into the lifestyle surrounding Mary Carson O'Hara of Pittsburgh is found in the journal of Mrs. William Foster, who wrote about the personality of people in Pittsburgh for fifty years. She was the mother of Stephen Collins Foster (1826–1864) of the *Old Folks at Home* fame. In Mrs. Foster's journal, General O'Hara and his wife are described as they appeared on the occasion of the marriage of their oldest son William in 1807 to his cousin Miss Carson.

> At General O'Hara's elegant mansion on Thursday, everybody was stirring. Polishing the windows and the massive silver plate was done by the butler and black servants. Old Hannah, the head cook, was preparing roasts, game, plum cakes, pound cake, loaf cakes, ices and jellies. At six o'clock the household's servants were clean and well dressed in clean calico and blue jeans while standing like statues. Everything was in order and gentility reigned.[1]

> [The] well proportioned figure of Mrs. O'Hara filled the elbow chair at the furthest extremity of the immense drawing room. Her large black eyes spoke of intelligence. Her deep toned voice breathed forth wit and eloquence. The queenly bearing of her person commanded homage from all who came within the sphere of her dignity.[2]

General James O'Hara, a true Irish gentleman was "dressed in the aristocratic garments of seventy-six; his locks frizzed and his long hair queued behind. A blue rounded dress coat, a crimson velvet vest and white breeches buckled at the knee over white silk stockings with gemmed buckles the same as those which fastened his highly polished shoes."[3] General O'Hara had four sons and two daughters and the youngest daughter, Mary, was about seven years old at the time.

Mrs. Foster described a pleasant evening where a large crowd had gathered to hear all the dance and music scholars of Miss Prevost perform.

Mary O'Hara was one of those scholars. Mary stepped forward, curtsied and took her seat and played the harp which was entwined with wreaths of fresh roses. "Her fingers swept the strings with such power in the more forcible passages and with such tenderness and correct accent in the gentler strains. She then sang clearly and sweetly without trembling or moving her features in wit and genious [sic]."[4]

She then danced a solo before she and Jane Wilkins played together on the piano, but her performance singing and playing the guitar was solo.

Again, after Mary had grown to be a young lady, she was the subject of Mrs. Foster's writing in her diary.

> There was a numerous company assembled. The hum of voices in conversation filled the room, when of a sudden every eye was turned towards where she was. At the request of her hostess she had seated herself at the harp and drawn her fingers over the chords. The weird sounds reached every ear in the two rooms startling them like the legendary horn of the Wild Huntsman on the charmed air of the Hartz Mountains. She paused a little to fix in her mind what she should perform, then swept the strings, for a moment loud, then faintly, as she commenced her song. The company were amazed. The song was sweet, powerful and enchanting as was her playing, but it was her beauty, her dignity of bearing, the graceful elegance of her movements, her intelligence and condescension of manner, produced a sensation that passed electrically around the rooms. And from one to another passed 'Who is she' and the answer, 'She is Miss O'Hara, she is from the west'.[5]

A letter from Mary Carson O'Hara (Philadelphia) to her sister, Elizabeth, dated, March 16 [no year] was written at a time when she was so tired of attending those vile tiresome tea parties and the invitations came daily. "I rode out yesterday with Miss Miller and the Smiths to the institution for the deaf and dumb. I never was more pleased in my life. I would rather go there than to the Theater."[6] She goes on to tell what a moving and exciting experience it was to learn how these children could read and write and she fell in love with a very intelligent handsome twelve year old boy which made it worth coming the thirty miles. Mary was looking forward to going home in May to Pittsburgh.

William Croghan, Jr. (courtesy of Historical Society of Western Pennsylvania)

Mary O'Hara Croghan (courtesy of Historical Society of Western Pennsylvania)

Since she was not to return to Pittsburgh until May, this letter must have been written before the year 1822, when William Croghan, Jr. met and fell in love with Mary.

Morrison Foster, brother of Stephen Collins Foster, wrote about the man that Mary O'Hara married who was the son of Major William Croghan from Kentucky.

> William Croghan, Jr., was a remarkably handsome man whom the writer remembers with pleasure for many kindnesses to him as a boy. He was tall and well built, with remarkably well proportioned features and an exceedingly keen and intelligent eye. He was a very Chesterfield in courtly manners, and a true gentleman in heart.[7]

Romance for William and Mary O'Hara must have bloomed instantly. Their wedding took place in the Presbyterian Church on January 23, 1823 in Pittsburgh, Pennsylvania. They were living in Pittsburgh in one of the O'Hara three story brick townhouses when their first child, William III, born March 1824, was baptized at Trinity Church, Pittsburgh, April 1, 1824. Eighteen months later, October 15, 1825, a second child Mary O'Hara Croghan was baptized at Trinity Church in Pittsburgh.[8]

At that time, William was in need of money. A letter was sent to his brother-in-law, George Hancock, who was a wealthy man in his own right, but cash flow was scarce and hard to come by. Hancock explains his inability to pay a debt at that time. He had tried to raise the funds, but without success however, he did send a check for $325.[9] William had expressed his dislike for farming, but for lack of income he was forced to return to Locust Grove, which he had inherited.

Spring of 1825 was when General Jackson, Rachel and family stopped off as overnight guests at Locust Grove while traveling from Washington to Nashville. The Croghan sisters, Ann and Eliza, had made quite an impression on him while in the capital city. Immediately after Jackson's visit, Lucy along with other ladies of Louisville was helping with entertainment for General Lafayette when he came to town.

Sadness again came into Lucy's life when one of her twenty-three year old twin sons, Nicholas, died in July or August of that year. Charles felt the loss more than anyone and was so distraught that family members had urged William to accompany grieving Charles to Washington.[10]

Charles, Eliza, Ann and their mother were together that winter in Washington. All would be in attendance for the social functions at the White House.

Back in Kentucky, Mary Carson O'Hara had left all this glamorous life behind in Pittsburgh and her life had changed dramatically. She now had the role of wife and mother in Kentucky, while the Croghan girls were far away from the farm life.

Locust Grove could have been a lonely, somber place while she presided as mistress, especially in the Fall. She must have longed for her friends and family back in Pennsylvania. Did Mary ever play the harp in the ballroom at Locust Grove or did William ever accompany her on the violin?

The Jesups received a letter dated June 13, 1826 from William, "My wife and children are well—I wish I could see you, Ann & little ones. Mary joins me in love to you."[11] He seems to be lonely and perhaps misses the presence of the large family who, only four years before, his parents, brothers and sisters, were at Locust Grove with friends and the house was ringing with laughter and happiness. John was down on the Cumberland River and Lucy was visiting Eliza at Fotheringay in Virginia.

By August 1, 1826, William was sick and tired of farming, incessant toil and anxiety and no profit: "I am firmly resolved to soon as my difficulties will allow to make arrangements for moving to Pittsburgh, unless the ponds between this and the river are drained, this place can not be healthy, the water in them now is 6 to 10 feet deep. My poor little son has the Bilious fever, he is better than he was although far from being well . . . The spirits of my wife are greatly depressed owing to the death of my little daughter & the subsequent illness of my son."[12]

William and Mary Carson lost their nine month old daughter. Mary O'Hara Croghan died July 18, 1826 and was buried in the family cemetery at Locust Grove.

A third child, Mary Elizabeth Croghan, was born 13th of April, 1827 at Locust Grove. When Mary Elizabeth was six months old, she and her three year old brother were motherless. William Croghan's beloved Mary died.

The date of Mary's death has erroneously been reported—even the date on her grave stone is wrong. The following letter would confirm the date as 25th of October, 1827 and it could not have been October 15, 1827.

My Dear Harmar, 11 A.M. 25 Oct.

I addressed you on yesterday informing you of Mary's alarming illness. It has pleased The Almighty to add one more day to the number of her days and I am permitted to say by her physician she is no worse, but yet dreadfully ill. The attending physician says he has hopes. They have partly reduced the fever. All they seem to fear is the want of strength in her system to heart. The palpitation of her heart and disposition so faint is dreadfully alarming. I must be candid with you, Eliz, & Dr. H., I think there is not much ground for hope, but this may in a great measure be the result of anxiety and distress. Everything that can be done for the dear creature shall be done and if it please the Almighty to take her to himself, we poor vain mortals must submit, Father, thy will be done. God continue to bless you & yrs. is the prayer of yours. W. Croghan.[14]

Lucy had retuned to Locust Grove sometime before August 28, 1827, as noted in a letter written by William to Jesup. At that time, William was not having an easy life and owing to this, Lucy had returned to be with her son and his family.

The first little Mary had been buried alongside her grandfather, Major Croghan, her famous great uncle General George Rogers Clark, and two Croghan uncles in the family cemetery. Mary Carson O'Hara Croghan was buried beside her little daughter in the family cemetery. Years later, both mother and daughter were re-interred in one grave at Cave Hill Cemetery, Louisville, Kentucky.

William honored his Mary's last request by taking her children back to her home in Pittsburgh. Immediately, following Mary's death, her sister, Elizabeth O'Hara Denny came on a sad journey to Locust Grove and accompanied William and the two children back to her family in Pittsburgh.[15] Soon after his return to Locust Grove, little son William III in Pittsburgh received a letter from his father [fall 1827], who had returned to Kentucky.[15]

My Dear Little son,

It won't be long now, before you see your Papa, who wants to see you and your little sister so badly. Can't you make Little Sister clap hands for Pa ... Little Abe & Al, find the most and Al would come in and say, "here old master here is egg, now give me cake" & then away he runs & then Abe, he comes in with his. Little Tommy & Susan live at the river, but they come up here of a Sunday to see us all." Little Harvey wanted to go with William to Pittsburgh and reminded him that he belonged son William III. Bob was living in town with the black barber that once cut his little son's hair and Bob was also, learning to be a barber. "...Kiss dear little sister for your Pa...Don't you say your prayers at night? Good bye my dear little Son. Yr. Father W. Croghan.[16]

Three months had past since Mary's death and the following letter was received by Ann and General Jesup from William dated, January 28, 1828.

> The dating this letter brings to my mind many gloomy associations in which if I were to suffer myself to indulge, I would soon be disqualified for my present purpose. But as it has been my constant effort since my great & greatest misfortune to struggle to keep up, when almost overpowered, & for which I have received my own & the approbation of my friends, it would not do now to give way. But, to think this day five years I was married & that now I am a poor dejected widower with two sweet little ones who look to me alone for comfort & support—the thought is very piercing, but I must submit "Father thy will be done, not mine..."[17]

William's life was completely shattered when three months later, his little son, William Croghan III died April 25, 1828 in Pittsburgh of measles and whooping cough.[18] William Jr. had taken great pride and satisfaction in knowing this son would be the fourth in line to carry the name William Croghan. Grief stricken, William became a thoroughly distraught and distressed man and must have hugged his little daughter closer since she was all that was left of his years with his beloved Mary.

Christ Episcopal Church, Louisville, was a haven where William would have gone to seek solace. He had been one of the founders of the church.[19]

The next year would be a difficult time of uncertainty in William's life. He was financially strapped and searching for direction. As the sole heir to one of James O'Hara's three survivors, his child would inherit a third of her grandfather's fortune and soon would be the wealthiest child in Pittsburgh. Later, the inherited estate would be worth millions.

No other one year old motherless child any where in America had been born with more patriotic blood in her veins than Mary Croghan. She was granddaughter of General James O'Hara and Major William Croghan, great niece of General George Rogers Clark, General William Clark and niece of Colonel George Croghan, hero of War of 1812.

In the meantime, William was floundering, since options were few. Despair had set in and things weren't working out in Pittsburgh as he had expected, due to the settlement of legal matters concerning his wife's inherited O'Hara estate. There was no working capital and Baldwin seriously advised William "... against having any thing to do with the Estate, for as Trustee, I would have to settle often with Court, if she [daughter] lived to be married (a thing probable, but yet a long while to look ahead) I might have difficulty with the husband ... so I have resolved to take Baldwins advice ... I sicken at the thought of farming in Kentucky. The truth is the want of occupation has rendered me good for nothing & my distress & disappointment destroy all my energies."[20]

He had borrowed a large sum of money in Louisville to rebuild the Crown Point Brewery in Pittsburgh as an investment which was part of his wife's estate. Having no ready cash coming in, he even decided to sell his valuable collection of paintings to a Mr. Frederick Rapp, Jr.

January 8, 1831—Pittsburgh, "... Many paintings as I have purchased. I never before offered to make a sale of any. Those paintings are of value beyond any thing I ask. I expect soon to leave here, & know not whether I shall again go to keeping house. I am therefore, anxious to dispose of those I offered you & others." Those he offered were:[21]

> 1. O'edipus Anigone & Polyneus (from a story in the Greek mythology) by Bilcocq, a French Artist. This painting was purchased in Paris, in 1814, by W. Lee, (a connoiseur) formerly our Consul at Bordeau & late 4th. Auditor of Accounts, & for which he then paid as he assured me 1,500 francs.
>
> 2. Is a head (doubtless and original) by Rembrandt, purchased at the same time & by the same person for 300 francs.
>
> 3. "The Head of a Miser" said by Van dyke (doubtful) certainly of fine execution.
>
> 4. St.Peter, by Tiepolo an Italian artist.
>
> 5. & 6. Concert, & Feast, Muller
>
> 6. & 7. Freebooters, & Money Changers. These are two splendid paintings & universally admired, by Vas.
>
> 8. "Blessings before meal." It has the appearance of being once a fine painting, it is much injured by age.

For those paintings, William was willing to take three hundred dollars and was sure they would be an acquisition to any gallery.

General O'Hara had placed his estate in a complicated trusteeship, making it difficult to take complete control. It was good that he would be living in Pittsburgh since he wanted to keep an eye on Mary's fortune, himself.

William applied his business sense and management experience he had acquired while managing Locust Grove and serving as an officer in the Louisville branch of the United States Bank.[22] He leased his wife's trust lands from the estate for $1000 a year, and got five or six times as much income from their use. He traded in stocks, buildings and lands and built up a considerable personal fortune in real estate which was about $25,000 beyond what was necessary to cover his debts.[23]

Before moving to Pittsburgh, William Croghan sold Locust Grove to his sister Eliza and husband, George Hancock. George Hancock had inherited his father's large estate in Virginia which included Fotheringay, slaves and hundreds of acres. Virginians had learned that planting tobacco, year after year, had a way of depleting the land of all nutrients necessary for healthy crops. There was almost no income from the Alleghany Turnpike since the steamboat was built.

George Hancock's mother relinquished her hold on the manor house estate and would live with George and Eliza at Locust Grove. Zachary Taylor's mother and George Hancock's mother were sisters. The Taylors lived on the adjoining property to Locust Grove and one of Mrs. Hancock's daughters, Caroline Hancock Preston, lived in Louisville so therefore it was not difficult for her to make the move to Kentucky.

Lucy left Locust Grove to live with Ann and family in Washington, but would be visiting her other children, from time to time. Dr. John Croghan had been living in Louisville since his father's death, but moved into town after selling Locust Grove to George Hancock and Eliza. George found the old home place in run-down condition. Fences had to be repaired and he was very displeased about bringing his well-trained industrious slaves in the midst of the slaves at Locust Grove. He had no desire to buy the slaves at Locust Grove and furthermore he had to invest $3000.00 in repairing and fixing the place up.

George Hancock did invest money in Locust Grove, bringing it up to a good working farm, and he lived there until 1834.

CHAPTER TWO

Mary Croghan Had Two Families

William's motherless child was fortunate to have a Grandmother O'Hara and Aunt Elizabeth Denny to care for her. The Denny household included parents, Elizabeth and Harmar Denny, and twelve children, eight of whom survived childhood. Mary Croghan would always refer to these cousins as brothers and sisters. Her Aunt Elizabeth O'Hara Denny would always be addressed as "Mother."

Harmar and Elizabeth Denny were community leaders in many civic and charitable activities. Harmar was active in the First Presbyterian Church throughout his life, serving as elder; Allegheny Bridge Company, manager; Eagle Fire Company, vice president; Secretary of Domestic manufactures in Allegheny County, Bank of Pittsburgh, director; Western Theological Seminary, director; President of board of trustees at Western University of Pennsylvania; Exchange Bank of Pittsburgh, director; Historical Society of Western Pennsylvania, president, and Allegheny Bridge Company, manager.[1]

Elizabeth Denny served as the anchor for the family, providing guidance and support to her children, grandchildren, nieces and nephews, long after they had reached adulthood. Although her husband and her sons-in-law often acted as her agents, she handled many of her own business affairs and was the only female stockholder in the Duquesne Inclined Plane Company.[2]

Mary Croghan must have taken notice of her Aunt Elizabeth's industry and leadership and her life was molded with the same attributions. The Croghan genes inherited from her father and Grandfather Croghan served her well. Mary was six months old when she had been thrust into a large family with several cousins. It can be argued that her early years were mixed with confusion—not knowing where she really belonged in a family unit. Her Denny, Jesup and Croghan cousins had one 'complete family.' She would have a divided family, even though she received love from all those around her.

In one of Mary's rare extended visits with her father, William Croghan wrote to cousin Lucy Jesup as if Mary was writing the letter, "When I visit my Papa, he is my father and when I go back to Mother and Denny, they are my mother and father."[3] Years later, when she was sent away to school in New York, her letters indicate what an impact this family situation had on her. Mary's need for reassurance that she truly belonged seemed to be missing and letters to her father reveal his lack of attention to her.

Mary was eleven years old when her grandmother, Lucy Croghan, died in 1838. During Mary's first eleven years, there were opportunities to be with Grandmother Lucy at William's home in. Pittsburgh. A Schenley granddaughter wrote in 1945 that as a young girl, her grandmother Mary Croghan had spent a great amount of time with the Jesups in Washington. Later, she and cousin, Lucy Jesup, were together for two years while at school in New York.

When young Mary was visiting cousins, William's time was spent in travel. In the beginning, this could have been a means of escape from living alone with memories. Very soon, it became apparent that he could afford a life style as a world traveler moving in the company of prominent and interesting people. At one time, many months passed before he returned to visit Mary.

William was a dashing handsome man about town and very wealthy. He wrote to Ann informing her that he was confined to his couch for three days as a result of leaping from his carriage in an effort to show his youthful activity to some young ladies and sprained his ankle. "I made a fist of it", said he.[4]

He was left at the altar by one fair damsel who had left town with his friend.

> ... Lieut. Phillips, who gave you the first information of my intended marriage to Miss O————y & expressed so much anxiety to be here in time to witness the nuptial ceremony, had the good fortune to lead her to Hymens Alter last week. Peace with them, I know not which of the precious couple to hold in the greater contempt.[5]

Mrs. O'Hara and Mary had taken residence at Mr. Denny's in the country, near the city. William's lifestyle left little time for Mary, as one can see from his schedule.

November, 1830—Mrs. O'Hara and little daughter were with William.

Janauary, 1831—William left Pittsburgh about 1st of February.

March, 1831—William was going to Naples, Italy.

In March 1833, William wrote to Ann Jesup that he had many plans for his spring and summer's campaign. He would stay in Pittsburgh until June—then to Louisville—then travel for six weeks playing "the man of business," and he hoped to spend time at Sulphur Springs for some few weeks.

Seven year old Mary was living with aunt, uncle and cousins when Ann received a letter from William in which he complained as to how much it grieved him to think he had not seen his little daughter for ten months. It seems as though Mary's Aunt Elizabeth Denny or her Aunt Ann Jesup had reminded him of this in a reproachful manner; however, in July he was in New York on his way to Canada, when the news of his sister Eliza's death reached him. He immediately canceled his trip and rushed to Washington to be with his mother. Dr. John was returning from Europe and upon reaching New York, he too joined William, Ann and their mother in Washington.

In September, Dr. John accompanied William to Pittsburgh for a short visit before returning to Kentucky. "... The Dr. was very impressed and quite well pleased with his brother's bachelor mode of life. Little Moll almost devoured her Uncle John, every other word being my Uncle John."[6] Mrs. O'Hara and granddaughter, Mary spent a few more days with William before he left for more far off places.

In the meantime, life in Pittsburgh went on for little Mary Croghan. The Foster family saw much of the Denny family, William Croghan, Jr. and daughter Mary during those years in Pittsburgh. Stephen's father William Barclay Foster had come to Pittsburgh, with his bride Eliza, as a business partner with Mary Croghan's Denny relatives and lived in the Denny household for some time.[7]

Stephen Collins Foster, was born on the 4th of July, 1826 and Mary Croghan was born nine months later, April 1827. Mary must have taken notice of Stephen's musical abilities when he was six years old or younger. He had taken up the flageolet and mastered the stops and sounds to the degree that he played "Hail Columbia" in perfect tune and accent.

Neighborhood boys formed an amateur "Thespians Society." Their theater was a carriage house and Stephen was the star performer. Soon, his leisure time was spent serenading under windows with friends.[8]

Many years later, the Fosters' son Morrison would always remember William Croghan's kindness to him and when Mary Croghan was an older lady living in England, she would be indebted to Morrison Foster for taking a firm stand against the state of Pennsylvania, making it possible for her to regain her inheritance.

In April 1833 William began building a mansion in the country which he named "Picnic House." This house was built as a partying place for his daughter, Mary. But, years later, to the citizens of Pittsburgh, it would always be known as the Schenley Mansion.

When Mary was older and living in England, she told stories of how she loved to run on the grounds at Picnic. Caretakers would later say, "The flowers were so thick that children playing had to be called from the wide wraparound porch. They couldn't be seen for the thick beds of roses and lilacs and lilies. Honeysuckle vines were everywhere—palm trees, too, and paths winding through the garden. The greenhouses were where they raised every exotic flower and where in winter were all the vegetables of summertime. Oranges and lemons were grown."[9]

William offered his cousin Charles W. Thruston, who served as his Louisville agent, an honorary membership in the Pittsburgh Horticultural Society. "We have Sundry public & private Green houses, boasting a vast variety of exotic, from far & near. We anticipate some public good & much pleasure from our association."[10]

In December 1840, Mary was twelve years old and going to school at Belmont Castle, New York. A Mrs. McLeod was proprietress and Mary's cousins, Mary O'Hara Denny, Lucy Ann Jesup and John O'Fallon's daughter were also students at this exclusive girls' school in New York. Daughters of other well-known families from Louisville attended this fashionable school and only the wealthy sent their daughters to Mrs. McLeod's school in New York.

At one time, Mary was expecting her father to visit in two weeks. She wrote him asking for a merino dress and one for Lib [cousin, Elizabeth Denny] for Christmas "a pair of side combs, some aprons, gloves and a breast pin. Pina, Lib and I have been talking about what fun we will have at Washington in March. I hope you were in earnest when you asked us if we would like to go for we have been seeing Mrs. McLeod about letting us go."[11]

In America, some families had become very rich. The industrial revolution, commerce, trade and textile industries in America made new millionaires whose family names would be recognized worldwide. For a young Englishman, especially a fortune hunter, "the thing to do" was to go to America and marry an heiress. A certain Englishman, Captain Edward Wyndham Harrington Schenley—cavalier in his early forties—had such a plan which was encouraged by his former sister-in-law, Mrs. Macleod, proprietress of the school in New York.[12]

When in his twenties, Schenley had fought at the battle of New Orleans, returned to his regiment, the English Royal Rifles in time to win a captaincy and a medal at Waterloo. He was friend of Percy Bysshe Shelley and Lord Byron. In 1822, he was with Byron and Shelley in Italy and according to legend, participated in the poet's funeral. Shelley's body, which had washed up on the shore of Viareggio, Italy, was cremated and the captain was supposedly one of those who took the heart from the funeral pyre. He had outlived two young wives and was father of a very young daughter. His first wife, Miss Inglis, died without family. His second wife, a Miss Pole, who was born into a family of English nobility, Powderham Castle near Exeter, died, leaving one child, Fanny Inglis Schenley.[13] Fanny lived with the Poles, which left him free to visit New York.

CHAPTER THREE

Romance That Rocked Two Continents

"Schenley Comes Courting for an American Heiress."

HEADLINES!! HEADLINES!!

February 1842

Wealthy heiress 14 years old elopes from an exclusive Staten Island School to marry a British officer, Edward W.H. Schenley, a man twice widowed and three times her own age, who was then a commissioner of the slave trade on a British Colony.

New-York Commercial Advertiser, Saturday, Feb. 12, 1842 "Much feeling has existed in certain circles, lately, occasioned of the clandestine marriage of a young lady—a *very* young lady—with an elderly English gentleman, who has for some time figured rather largely in our fashionable quarters. The young lady, Miss Croghan, niece of Col. Croghan, was at a fashionable boarding school on Staten Island, kept by a lady whose daughter is said to have been the first wife of the elderly English Gentleman. Miss Croghan is said to be only 15 or 16 [she was 14—born Aprl 13, 1827] years of age, and heiress in her own right to a very large estate. The gentleman, a Mr. Shenley is variously represented as to age, from 45–60, but a skilful "maker-up" so as to appear much younger, and of very seductive manners. The marriage, it is said, was performed by one of the magistrates—incautiously, no doubt, but without connivance at the secrecy observed. Immediately after the ceremony the parties embarked for England.

The father of the girl is said to be, and no doubt truly, very afflicted. It is possible, however, that he is relieved from the care of an undutiful daughter. One thing is obvious enough—that such an occurrence is by no means extraordinary, as a result of education at a 'fashionable boarding school.' "

Mary Croghan Schenley, age 14, from a newpaper drawing 1842 (courtesy of Carnegie Library of Pittsburgh.

New-York Commercial Advertiser, Feb. 17, 1842 "We are happy to learn from authentic sources, that Captain Wyhndam *(sic)* Schinley *(sic)* who was recently clandestinely married, with the co-operation of a son of the mistress of a fashionable boarding school at New Brighton, to Miss Croghan, the niece of Colonel Croghan and General Jesup of the army, will be defeated in his principal object.

We are informed that for ten years to come the present Mrs. Schenley can have no control over the large estates devised to her by her grandfather, the late General O'Hara, of Pittsburg, but is entitled by his will to an allowance of only one thousand dollars per annum, and further, that the consent of her father to her marriage is necessary before the property vests in the young lady at all. We shall be extremely happy if other individuals implicated in this lamentable affair clear themselves from the charge of connivance of participation in proceedings which have excited so general a feeling of indignation in this community."

Edward Schenley (courtesy of the photographic archives of the Historical Society of Western Pennsylvania)

New-York Commercial Advertiser, Feb. 22, 1842 "The late elope-ment. *The Pittsburgh American* sets forth the manner in which Captain Schenley will be frustrated in his designs upon the fortune of Miss Croghan. Her grandfather's estate is vested by will in four trustees, of high character, whose authority is absolute, touching the division of the state among the heirs, of whom there are three—Miss Croghan's mother being one. Of course they will take care that Captain Schenley shall derive no benefit from the wealth at which he supposed he was making such a successful grasp.

The following letter is very interesting. It was written by Mary to her father, January 23rd, one day after she was married on the 22nd.

Tomkinsville, N.Y.

January 23rd, 1842

My dear father,

I think you have treated me very very badly indeed in not writing to me as soon as you arrived in Washington. If Emmeline had not writ-ten to Mrs. Macleod something about your being there, I can not say all the things I would have imagined had happened to you, but Never mind if I do not receive a letter from you tomorrow or next day, I will write another to you. Mr. Schenley has not yet ceased in his kindness to me and all of the other girls. The Saturday after you left, I went into the city with Mrs. Macleod to have my teeth [or tooth] arranged. After we had finished "he" came and took us to see Stouts statue of Fanny Elsler (Oh, it was perfect) and afterwards, we went to see the Panorama of Thebes and Jerusalem, that was quite enough for that day, and last Friday evening he took Fanny Wash, Mrs. Macleod, Pina and me to the theatre, we staid at the American (tell Emmeline we had the same rooms exactly) we saw "London Assurance" over again and "What Will The World Say" O! It was too too nice. I like the last the most, as it was very very amusing and interesting. We had the same private box that we had the first night.

I wrote to Lutie [Lucy Ann Jesup] directly after she left here and she has not yet honored me with an answer. I have been almost tempted to try Emmeline and see if she has yet forgotten old (& young) friends, but something always says to me "not yet, not yet" so I will wait a "little" longer ... I want to get a cloak and bonnet, two very necessary articles for New Brighton, and I thought it would be better to tell you I want them before I get them, am I not an excellent good "big" girl I think so? Do you intend visiting New York before you go to Pittsburg, from what Emmeline said in her last letter you had not decided—Good bye my dear pa—if you do not soon write to your very affectionate daughter Mary.[1]

One wonders if that last line, *Good bye my dear pa*, had a special underlying significance—also, not yet, not yet. Had she been laboring over making the decision to elope and waiting for her father to show some signs of giving more love and attention? If not, a charming and seductive "father figure" was waiting in the wings, ready to take her father's place. In reading family letters, it is easy to assume that Schenley and his first wife's sister, Mrs. Macleod, never failed to make use of every opportunity to point out, in a subtle way, the "shortcomings" of her father and at the same time, Schenley appearing to be her knight in shining armor, who had come to rescue his princess. After all, it was Macleod's letter (from Emmeline) that was shown to Mary in reference to her father [who had failed to write] and how he was spending his time.

While this elopement was taking place in New York, William was visiting the Jesups in Washington D.C. This was the place to be for the holiday social season. The big social event taking place at the White House for that holiday season was the wedding of President Tyler's daughter, Elizabeth. She was married in the East Room in January 1842. Dolly Madison was there. The Jesups were always present for these events and members of the Croghan family would have been included. William's brother-in-law, Denny, was there.

After the elopement, it would be two weeks before the family received the news. William was still with the Jesups when he heard of his daughter's marriage and the news came in a letter from Schenley's long time friend, Henry Delafield, New York, dated February 2, 1842. He wrote, "Your surprise on the receipt of the enclosed communication cannot exceed that which I experience on the receipt of this information."[2] He was also surprised that he was made the "go between" of communications for Schenley and William. Since he is at a loss on how to proceed, he sent a copy of Schenley's letter, giving details and particulars. Nothing he said could console the father of this painful occurrence, but Delafield did add that Schenley had a high position in Society as that would soothe the pain. He did not know whether Mr. Schenley sailed in the packet of yesterday or not, but presumed he did. For some time past, he had not seen Schenley.

Slow communication in those days, aided in the success of secrecy. Newspapermen from St. Louis, Pittsburgh, Washington D.C., New York— all were writing front page articles about the "kidnapping."

> The young girl's father had a stroke or fainted upon hearing the news ... The U.S. Government had sent a ship to try and capture the lovers on the high seas.[3]

These rumors lasted for years. Perhaps, upon hearing the news, William did go limp and collapsed in the nearest chair, but a ship sent by the government would have amounted to war or piracy on the high seas.

General Jesup immediately went to New York to investigate the unbelievable news in order to find Schenley and confront Mrs. Macleod. William wasn't up to this job. Jesup's daughter, Lucy Ann, was certainly not to return to this school.

A letter, dated Feb. 16, 1842, from Colonel George Croghan, Mammoth Cave, to cousin John O'Fallon in St. Louis, made very clear his feelings on the matter;

> My Dear sir;
>
> I entreat you as a friend and relation to take your daughter away from the Brighton school as you would save her from the contaminating influence of its Directress Mrs. McLeod, than whom a more artful, intriguing and base woman does not exist. You will have heard that my Brother William has been robbed of his Daughter, a child of fourteen years old—Yes, saddened by the wheedling artifices of that vile woman she has eloped "[indecipherable word]" away & sailed for England with a Mr. Schenley, a man of fifty-six [?]and brother-in-law to Mrs. McLeod who has for a length of time been aiding and abetting with fiend-like appetite his worthy accomplice in the nefarious scheme of robbing a father of his child, that they may secure to themselves a portion of her vast estate
>
> Mrs. McLeod will attempt to exonerate herself from all blame and may succeed with some for she has the talents & deceit of the devil himself, but listen not to her. Facts are so strong against her that nothing ought to restrain my Brother from arraigning him before the courts as the kidnapper of his child.
>
> Two years ago, if not more, the Dr. implored Wm. not to enter his Daughter to the care of Mrs. McLeod as she was very unworthy— had the Dr. prevailed what agony would have been avoided.

I have received two letters from Mrs. Croghan upon this distressing subject filled with details of the cool, calculating scheme and artifices resorted to by the vile woman to affect her nefarious end. Wm. is half distracted. Let him rouse himself and pursue to the rescue of his child, even though to affect it he will have to blow the vile robbers brains out. I write in haste and in great distress. My love to your family Yours affectionately, G. Croghan.[4]

George Croghan has to be admired for his quick, concise solutions for any grave issues at hand. Schenley's daughter was only two years younger than Mary Croghan and her father William Croghan was only two years older than Schenley. John wrote to Jesups, 9 May, 1842 ... "William arrived here last week and spent a day or two with us. The only topic (almost) during his stay, was the abduction of Mary."[5] Once more, William's life was shattered and knowing not, which way to turn for the future.

After William learned of his daughter's marriage, he asked the Pennsylvania Legislature to pass a law, whereby, all of Mary's trust funds should be held by him. The Pennsylvania State Legislature met to make sure that Mary's father, William, retained control of the money. One state legislator declared: "No English officer, much less the fortune-seeking Schenley, is worth one hair on the pretty head of an American girl."[6] William Croghan, Jr. was made an arbiter of Pittsburgh, Pennsylvania and The Staten Island School broke up over the scandal.

In thirty-six years of marriage, Mary Schenley never took control of the O'Hara-Croghan fortune. By law, she could have liquidated the holdings upon reaching the age of twenty one, but, by her direction the property remained under the control of a trust governed by Pittsburgh attorneys and bankers. By that time, Mary realized her husband had his weaknesses. Only after her husband died in 1878 did she make preparations to dissolve the trust and sell or give away the considerable holdings she inherited, everything except Picnic House.

Selling such a large block of property without suffering a loss was a task that would employ the trustees for nearly forty years. Eventually, Picnic House was the last relic of the O'Hara-Croghan, Schenley line and was occupied only by caretakers for about sixty years. The mansion fell victim to the wrecking ball in 1945 to make way for a new housing project.

During her father's first visit to see Mary in 1845, it is possible that he advised her on the subject of her estate. At that time, Mary would have been only seventeen years of age.

CHAPTER FOUR

Surinam—Dutch Guiana

Upon the bride and groom's arrival in England, Schenley's second wife's family, the Poles, grandparents of his daughter, who under trying circumstances, opened their doors and hearts to Mary in a distant land. Family letters indicate, it was Schenley, not the young bride, with whom they were not pleased.

Schenley and his new wife wanted to remain in England but were sent to Dutch Guiana (Surinam) where he was a slave-trade commissioner.

In the new foreign and strange land, Mary, feeling forlorn and forsaken, longed for America, her father, cousins and friends with an aching heart. The following letter was sent to her father after seven months away from home and only one letter from him.

Paramaribo

3rd September 1842

My dearest Father:

I can no longer be patient; Seven months have I been away from home; & have but one letter from my dear father; I have prayed to be contented & say "it is all for the best,... But still no letter; ... oh pray tell me, why will you not write to me? You may be ill & I not know, you may be well & enjoying yourself & not thinking of coming near me your only and devoted child. If I have done wrong my dear dearest Father forgive, I am wretched for what I have done to you & the only manner & my last ray of hope that I have of comforting you and myself, is cut off. By your being persuaded not to come near me. If Mother & Denny tell you what they think best, I am confidant they will advise your meeting me & my dearest Husband; oh my father it is for you I am now thinking; not for myself . . . No no not for myself . . . Oh why do you

not come; you will at once in this climate; be restored to the enjoy-
ment of your health; & spirits I know; and oh I assure you what
a good child I shall be, (with "God's help") I will you repay all I
have done to you. But oh what a climate this would be for you.
Do do come . . . We ride every evening I read, practice, and do
other things to improve my time but oh if you would come only
this winter. Oh why wont you to make yourself well. Mr. Schenley
is still what he has always been a devoted, kind, affectionate &
every thing that's good Husband. Oh dear dear if you only knew
him. Please do write to me & say you will come. I only write to
let you know I am well, as happy as I can be without you & to
pray & entreat you to come to me soon. Oh do do do write to me
& put it in the mail at Pittsburgh "to the care of the Dutch Consul
at Boston & I will be sure to get it. Good bye . . . My dearest
Father, believe me ever your much attached and devoted child.
Mary Schenley.[1]

July 1, 1843, Edward Schenley wrote to Elizabeth Denny, "Dear
Madam, Your Dear Daughter was confined happily yesterday morning of
a very healthy child . . . daughter . . . both are doing well." July 9, 1843,
Schenley makes a favorable report on the ninth day of Mary's confinement,
"I enclose for the satisfaction of her family Duplicates of the legal certifi-
cates of birth executed as formally as possible." He stated that he did not
believe the child to be a native of the country of her birth, but an
'Englishwoman' because she was born under the English flag.[2]

Schenley's love for Mary appears to be genuine. He was protective and
concerned about her health during her pregnancy. After the birth, he sat by
her bed and wouldn't allow many visitors at one time and for long periods
of time. Schenley and Mary looked for every vessel in anticipation of her
"Mother" and some of the Denny family arriving. Even when Mary was
recovering form childbirth, Schenley was still writing to Elizabeth Denny
and urging her to visit. Mary had so much love for her and the only "parent" she now had. Schenley could not understand why "he" [William
Croghan] would not forgive inasmuch as it seemed to Schenley that he,
himself, would want to contribute to her happiness.

Elizabeth Denny, for some reason, was in possession of the letter which
Mary had sent to her father, September 1842. Mary was pleading for her
father's forgiveness in the letter. This letter must have been sent back to

Mary (by Elizabeth Denny) with William's reaction or reply, which would have been hurtful to Mary. The letter was intercepted by Schenley, who returned the letter to Mary's Aunt Elizabeth Denny with the following message, "I take the liberty of returning a letter that of right is your property, but which you for; doubtless; good reasons, enclosed to Mary. I consider it in July that a daughter whose anxious desire is to love, nay; venerate her only parent should remain in the possession your document that manifests so much reluctance and annoyance at having been given to understand that he might contribute to her happiness. I trust that, here, all such subjects may end. Believe me, Mary's mention of this was entirely without my knowledge or suspicion."[3]

In August, Mary wrote to "My Dearest mother" in which she wished her mother would come to see her and dearest little baby which was such a dear little plaything and was her own. She was very disappointed at not seeing her mother on the last vessel and would not be in Pittsburgh for a very great while. Schenley would be going to England the next spring and, if he goes there, she would not stay behind, but only if her Father could be persuaded to *meet* her in England. She could not ask *him* to, and she need not repeat any determination not to quit him [Schenley].[4]

Her new home in Surinam, a British territory in North Guyana, was quite different from the sheltered lifestyle she had left. It was a great awakening for her, especially when exposed to the world of slavery. She was only six months old when she left her birthplace, Locust Grove in Kentucky.

The following gives insight into who Mary Schenley was and why she stands out as a special person, even at the age of sixteen. Her perception, sensitivity and understanding is exposed when faced with a situation in her own household which she was unable to accept.

Mary would not be going away, until her baby was nine months old. She had a very nice and healthy Curacao Negress for her wet nurse and her baby was very well. Mary had discovered that great proof of the woman's health was her own child, which was eight or nine weeks older than Lilly, but was such a little fatty.

The difference between her child and the little black one had caught her attention and was of great concern. The other child never had a stitch of clothes upon it and lies all day on its back upon a leather [unintelligible] upon the sand in the yard, beside its mother's washing tub. If it rained, she was obliged to be in the house.

She ties her child onto her back by means of a long strip of muslin. The child's little woolly head, all that is seen and it's legs are stretched one on each side of it's mother. Here, it will be quite still and if it cries, the ma sticks her finger into mouth and say "tah monso, in bukia hous" —"shut your mouth in white man's house". This, generally makes it stop young as it is, if not it is sent into the negro house to cry itself to sleep.

But, while it's in its 'seat of honor' upon it's mother's back, she nurses Lilly and does her work exactly if nothing was there. She is now teaching it to *sit down* or as we say, sit up alone. She puts the baby in a large tub and surrounds it with clothes in such a way that the little thing can not move itself and is sitting upright for a considerable time during the day.

But, to tell you any thing about the poor negroes, is too much and too ridiculous. At present, Mr. Schenley is very much occupied in endeavoring to procure for some hundreds, their freedom which for 20 years has been kept from them by the Dutch *and called free.* Their tickets of emancipation, never having been delivered to them, (owing to the negligence of the former British Commissioners) then of course, were treated as slaves and many, many have died during the 20 years. Mr. Schenley has been doing his 'duty in that state of life in which it has pleased God to call' him, and has obliged the Governor to deliver one cargo that were emancipated, their 'tickets', which has rendered them as free as I am and before long, he will succeed in releasing many others.[4]

A very interesting comment in the letter was: "I think Mr. Schenley is the 2nd 'great man, Washington.' "[5] At that time, she had not heard from her father since he had met with the accident of being thrown from his carriage and was concerned.

There was no controlling Mary's outspoken word at this grossly offensive and shameful state of affairs such as the slave situation. It was not with a reckless disregard for the consequences that she stirred up quite a bit of trouble for her husband. Mary's husband was in a position which gave him a voice in protesting and serving as a mouthpiece against this treatment of negroes.

"April 13th [1844]—Today is my seventeenth birthday," Mary wrote to

Lib [?]. Schenley had also written a letter to Queen Victoria about the slave conditions in Surinam and one planter was convinced Schenley presented a one-sided picture of an incident. This developed into a situation that aroused public attention and reached a heated level back in England. The Schenleys had made themselves unwelcome and were recalled to England.

It would be some time, if ever, before Mary would be accepted in society. Queen Victoria refused to have her presented at court because she had been "a disobedient daughter."[6] A letter sent to her cousin, Elizabeth Denny, Pittsburgh, in January 1846 gave an indication that something didn't go right for Mary in celebrating this event.

The Marquess of Anglesey had been kind enough to give Schenley a ticket to admit Mary to the House of Parliament to hear the queen's speech on the 22nd, but it came too late for her to attend. Since she had never seen a procession such as it was to be, she "preferred taking the two children to see the queen and her cortege as they proceeded on their way to the House. It was really a beautiful sight and we had a corner room in the hotel directly opposite Westminster Abbey when they all passed."[7]

The queen was in her state carriage drawn by eight horses. She was dressed in a white watered silk, wore pearls, and her hair was in a broad plait in front. Mary and the two oldest children were so near they saw everything of interest.

Mary wrote a letter some time later, in which she opened up her true feelings about being so lonely and sad. It would be impossible for her to visit Pittsburgh soon in reality, but was often there through fancy; unshackled by earth, and she "shall direct her wings to her dear friends and bring back those happy days they had spent together." She also realized fancy was but an idle vision even at best and hope would "lend her golden wings for something more substantial. I am getting into the clouds and must return to earth and reality."[8]

Oh, how she must have longed for her home and friends far across the ocean. Her young carefree days had ended much too soon.

Summer 1845, William had received a letter informing him that his daughter was ill and immediately found himself on a voyage across the Atlantic to be with her. This decision was made with very mixed feelings and emotions, he wanted to see his daughter so badly, and with the urging of his wife's sister, Elizabeth, he was to deal with the hatred he had for that son-in-law.

The following could be called *"William's Garden of Gethsemane."*

"Shute House" [England] Oct. 1, 1845

My Dear sister,

I presume you received the letter I addressed to you to go by the Steamer of the… & am surprised to see that I am still the guest of Sir W. Pole. Mary was anxious, I should defer my visit to London, until they took up their quarters there & be their guest. Our arrangement altogether agreeable to me, & possession of their house they could not get before the 6th Inst. So we leave their most enchanting hospitable place & it's agreeable inmates on the 3rd for Southampton, where we sojourn 2 days & then proceed to London, where I will look about in for a week or two, then for Edinburgh, Dublin, Paris & home—From the time I left you, my dear sister, until I arrived in the neighborhood of this place, I could not realize it. I was making my way to the presence of my child, or that I again to see her—every effort & movement was purely mechanical. I had it all "by heart." I was to reach Liverpool the best way I could & then be governed by circumstances on my arrival there. I received a letter from Sir W. Pole [father of Schenley's last wife], inviting me to his house & pointing out the route etc.: by cars to Birmingham, Bristol, & the station at Taunton—on making known my wishes to go to "Shute" I had a dozen applicants, all familiar with the route etc. etc, I was soon "porting" on the way, but instead of immediately going to Shute which was equally near, I chose to go to Westmister [?] to compose myself for the occasion & draw a long breath—how bewildered I felt, how agitated, with what conflicting feelings I had to contend, during my brief sojourn in the little town, I cant express—I realized then most fully and sensibly what I was about; in the fullness of my heart I could but weep, I fell upon my knees & most fervently prayed to our heavenly Father, to take me under his special care, to guide me & to direct me, & to purify my heart of all unkind or unchristian feelings— I think, I told you how everything went off on my arrival here, & I felt assured I acted on the occasion, as you would have wished me, but not expected of me. I am pleased at the course I pursued, for many, many reasons, not the least, the gratification it has afforded my poor dear child—my intercourse with him is unrestrained, very friendly & as prudence & policy points out the way, I hope my thoughts may never recur and past transactions that so sorely tried me & embittered my life—he is verily I believe attached & devoted to her & makes her the kindest husband, & equally devoted to him does she seem; Their two dear little children engross their thoughts—that is not to be wondered at, for two more lovely children I never saw & as often as I contemplate the happy group, Father, Mother & children in happy intercourse I feel subdued & from my heart silently ejaculate—"God Spared Me." One of my apprehensions was, she was a mere instrument in his hands, never thinking or acting for herself. Moreover from her extreme youth, she was not equal to the important trust, of training up as she ought her offspring ets. In the first place, I assure you he counsels her as a husband should his wife & she is as independent in the premises as it is discrete a wife should be. But, above all, her management & deportment toward her children really delights me. They are not humored or spoiled as I apprehended. She is as precise & strict as a Mother, I conceive, should be—If he had his way, he would soon spoil them, She frequently admonishes him, not to do this or that to them & he acquiesces. Her turn is truly domestic, she is industrious; while sitting, talking to me she is never without some work in her hands, making as she does all her children's clothing & assures me, she has no disposition to extravagances.

Never have I experienced more kindness & attention than I have, since my arrival here, not alone from this most charming family but from others, reached by their great influence. I reserve until I return my account of that most delightful visit.

Yr. Aff Brother

William Croghan

... let me know how everything goes on at Pic Nic—direct your letters to "Windham Club", St. James Square, London..... [9]

Schenley's demeanor was quite different from what he had expected. His son-in-law [two years younger than William] was patient and almost as one of the children and Mary was the decision maker and disciplinarian. She loved him, but letters reveal an awareness of his weaknesses. The rest of Mary's life would be a marriage to a husband who seemed to be in limbo, waiting for his wife's fortune to come under his control, but he was a devoted husband and father.

Wonder where Schenley got the idea for his devotional exercises at dinner? Was it a role he was playing, knowing that Mary had grown up in the Denny household where, before each meal, reading scriptures and giving thanks for their blessings was given. William wrote about the first meal he had at the Schenley's. Mary's husband made a big performance thing out of standing at the head of the table and reading with family worship, "enacting the part of Chaplain with becoming fervor & gravity. Before he partakes of his meals, he asks a blessing & at conclusion returns thanks,—he not only regularly attends Church himself, but all over whom he has influence or control, makes or likewise." William evidently was not taken in by his show of piety that was put on to impress his father-in-law, and no doubt, with all the drama of a two-penny opera, fell short of its aim. William wrote to his sister, Elizabeth Denny, "I see [unintelligible] beyond the mere surface."[10]

As time passed, and William spent more time with his daughter's family, he learned to accept his son-in-law. It appears as if William was saying ... "I know you are a rascal, but an interesting and charming rascal that I can tolerate". To the end, he thought Schenley was in debt and gambled.

Within a few days following William's arrival, the Schenleys and William moved from London to a small beautiful town, eighty-four miles away and a two hour travel in the car which left several times a day. The house was large, elegant and furnished handsomely with all the appointments. There was a picturesque view all around and from the drawing room was seen a large expanse of water, with vessels sailing to and fro. In the distance on the opposite shore were beautiful white cottages and many spires. The terraced garden extended along shore and the balcony extended the whole distance of the house which commanded the most beautiful view.[11]

There are some indications that William had purchased this beautiful estate for the Schenleys after they returned to England from Surinam and had very little income. The allowance they had been receiving had been increased.

William was enjoying very much his time in England. He dined at the *Rexbridge House* in London as guest of the *Marquis of Anglesey*. The

Marquis had also served in command at Waterloo. Mary's aunt Elizabeth ["Mother"] in Pittsburgh, received a letter from William in which he wrote, it was hard to believe, "... this plain unsophisticated, matter of fact brother of yours, was seated at dinner with Marchioness on one side and the beautiful daughter, Adeline on the other."[12]

Mary made William proud and happy when she named his new grandson "William Croghan." He was thriving and doing well. The grandfather had a hard time doing justice to the other two "Lovely little creatures, I have rarely ever seen." The oldest, he thought, was the loveliest of all. She would make her appearance in the drawing room each morning with her Mama and without prompting would salute with, "'Good morning Grandpapa." At Prayers she knelt down and never spoke or moved. Nothing seemed to escape her and she had a wonderful memory. Jane, on the other hand, was a good natured, fat, hearty little thing that talks a lot and delights in playing hide and go seek with Grandpapa, "... sad work do we sometimes make in the parlour by upsetting chairs." Mary and Schenley insisted Jane resembled him, but he couldn't see it and thought she was a homely likeness of Mr. Ross Wilkins and had told Mary.[13]

While in London, William found Watkins, the youngster that had lived with him in Pittsburgh and had attempted miniature painting. Watkins had been residing in London for ten years with his young family and was enjoying quite a high reputation as a miniature painter. William had become very busy in having miniatures painted of himself to send others. Found on the back of William's miniatures are the initials, "W C." Miniatures were much in demand at that time and it was a sign of cordiality to send loved ones a miniature of yourself. He insisted on miniatures painted of Mary and granddaughters.[14]

William wrote to Elizabeth that he, Mary, and Schenley too, ("who, to give the devil his due, as the saying is") would often express their wish that she was there with them. He also went into detail about the time he was spending in the London shops which to him were the most magnificent shops on the globe and purchased rare findings that only a millionaire could afford.

His letters contain very interesting reading about the many prisons he visited while in London and hoping to learn more about their prison systems. One letter revealed more than words can describe how his heart was touched by the less fortunate.

The last prison he visited, there were fifty little boys standing so as to make a hollow square, each with a book in his hand, the testament, and not daring to look up. Their ages were from nine to fifteen and all had been guilty of crimes and undergoing their sentence—six hours to labor and four to study. After reading, they would then close the books and answer questions. William was surprised at how proficient they seemed to be, since most had not known a letter in the alphabet.

He told the boys that he was from a distant land, but thanks to the Christian spirit that was within him. "While making those remarks or something to that effect, whilst the poor little creatures gazed at me with great intensity, I involuntarily drew near to me a poor little sallow faced fellow & patted him affectionately on the head. Poor little creature, whether it was my business, or what I won't say, he burst into tears & in which others joined. I need not tell you simple soul as I am, I joined likewise & not far off of it was their kind copious instructor."[15]

William came home soon after writing that letter. If not, he returned upon hearing the news of his sister Ann's death in April and after arriving, stayed some time with the Jesups before going to Pittsburgh. In June, Ann's children were on their way for an extended visit with their Uncle William at Pic Nic. Mary Jesup had married in January and since she and husband James Blair were expecting their first child, she stayed in Washington while James escorted the other three Jesup sisters and brother Charlie to Pittsburgh. He returned to Mary after two days at Pic Nic.

On January 1, 1847, William Croghan once more boarded the Good Steamer Cambria bound for England. He waved "good bye" to his nieces, Lucy and Jane Jesup. It had been eight months since Ann had died and since that time, William had been closely attached to her children, giving them love and comfort. That was a reverse role for William, playing the "father role" for Jesup's children, since he had neglected to do so for his own daughter Mary when she needed him. Evidently, Uncle William and Ann's children had pet names for each other.

Arriving in the British Isles, he spent more time in Ireland and Scotland. Much time was also spent in Edinburgh and long would he have occasion to remember his visit to that classy picturesque city. He would fondly remember the kinship and their hospitality and affection extended to him by his relatives, the O'Haras. Their recognition of him, he had little expected. He was sick in his hotel room and an O'Hara gentleman showed up, manifesting for William the most intense interest and anxiety and as soon as he was able to be removed, the relative would listen to nothing but bringing his carriage to the door and William was transferred to his house where the family was so attentive to him. William wrote to Elizabeth Denny, "This recognition of me as a relative, I little anticipated and was much affected by it. With the character of my poor dear wife, they were perfectly familiar and by no means ignorant of my history."[16]

Mary O'Hara's goodness, beauty and many talents were remembered by her Irish relatives. Knowledge of William's prominent role in society and his business success in America had preceded him to the old country.

Mary and her new infant Henrietta Agnes were both doing well, but at Mary's request, William remained in England longer since they were waiting anxiously to learn what the Foreign Office meant to do with or for Schenley. While there, William's brother-in-law Harmar Denny thought it necessary for William and Mary to negotiate and arrange matters in relation to the estate. Mary and Mr. Schenley were most willing to confer unlimited authority in the matter upon William, but what would it avail her being a minor. William thought he, himself, should assume the power and leave it to her on becoming of age to confirm it.

When William Croghan returned home, he had every hope of Mary returning with her family and making Pittsburgh their home. He immediately started enlarging Pic Nic, duplicating their home in England including the wide porch that wrapped itself around the house on three sides with its white colonnades and an English basement. He planted an English garden in an effort to lure Captain Schenley to Pittsburgh. The rambling twenty-nine room red brick house set on a one-hundred-five acre tract was believed to be the largest single piece of property owned by a single family in Pittsburgh. It was one of the finest examples of Grecian Revival architecture in America.

Mary did return from abroad with her English husband and many English servants when she turned twenty-one in 1848. She was expecting to claim her inheritance. By law, Mary could have liquidated the holdings upon reaching the age of twenty-one, but, by her direction, the property remained under the control of a trust governed by Pittsburgh attorneys and bankers. She did get a large part of the estate in August, and just in time, because two months later, the English Government retired Captain Schenley on a mere £1000 a year. The Schenleys collected enough income from rentals on the property to live very well.

Mary was very astute in making the decision for Pittsburgh attorneys and bankers to control the trust. Some of those attorneys and bankers were related to Mary on her mother's side.

Very soon, Schenley was tired of Pittsburgh. He felt as if he had been living in the backwoods of the world for the past year and it wasn't exciting enough for him. The following February, the Schenley family was going back to England which was a great disappointment to William.

William wrote to his "sister" Elizabeth, expressing grief over suggesting Mary should stay with him at Pic Nic since this must have presented a great conflict for Mary. Her heart was with her father and wishing to make him happy, but on the other hand, leaving Schenley on his own could lead to problems. William received a telegraphic dispatch from Schenley in New York where the family was waiting to embark for England. It had been decided that Mary was to return and spend the winter with her father. William regretted he had insisted she stay with him in Pittsburgh. Not wanting her to leave, he felt the poor dear had been willing to sacrifice herself just to gratify him and he was ashamed and humbled. He would rejoice to have them all with him and would assert himself to the utmost to render their time pleasant. "...But knowing what a sacrifice they are about to make merely to gratify me, I tried to discard all selfishness and act with civility and communicated by Telegraph a message to this affect." He told them that he was reconciled to their return to England and not to act on his feelings.[17]

He reminded his sister how sore his heart had been ever since his last return home and how greatly distressed he had been, but yet, he did not want to complain or say anything to distress his poor child or wound her feelings. In contemplation of her leaving, it had been a task to write to her or Mr. Schenley. His feelings would get the better of him," but what parent that had a heart could [unintelligible] contemplate the departure of his beloved and his only child for so distant a land." His heart yearned to see his dear child and those sweet little ones. He did not want to interfere, "but let them decide for themselves and so be it in the name of the 'Almighty'." "Pic Nic October 11, 1848." He had walked out on the terrace for he dreaded to go to bed for fear that sleep would not come to him.

Two months later, after Mary left her father, he received news of the deaths of his last surviving brothers. Inspector General George Croghan and Doctor John Croghan died within three days of each other. In July 1849, Col. George's body was returned to Louisville to be buried in the family cemetery at Locust Grove. A funeral service was held at a downtown church and William, George's children and many cousins must have returned to Locust Grove when George's body was laid to rest with his family at Locust Grove William was alone.

The Pittsburgh Morning Post
Pittsburgh, Tuesday, September 24, 1850

> The death of William Croghan Esq., on Sunday evening, the 22nd instant, at half past 7 o'clock, was sudden and unlooked for, and his confinement was but of a few days. Mr. Croghan was one of our most esteemed and wealthy citizens, and his death, unless he has otherwise willed, will greatly increase the already princely estate of his daughter and only child, Mrs. Schenley, whose inheritance came from her grandfather, the late Gen. O'Hara.

> The funeral will take place this (Tuesday) morning at 10 o'clock, from his late residence. Carriages will start at 9 o'clock, from the residence of Harmar Denny, on Third Street.

William's life had been filled with much heartache and loneliness. In the end, as they had been before, memories were his constant companion. He was left to live without his Mary and their children.

William would always lie in a grave in Mary O'Hara's beloved Pittsburgh. Mary Carson O'Hara's grave would remain in William's Kentucky.

CHAPTER FIVE

Schenley Children

When the Schenley children were very young, a seventeen year old girl, born in Middlesex, England, was hired and stayed with the family for sixty years as nurse to the nine children. The young girl married, becoming Mrs. Elizabeth Koehler and had nine children of her own.

During the sixty years with the Schenleys, Mrs. Elizabeth Koehler made five trips across the ocean. First, she was nurse to the Schenley children, then she and her husband lived on the grounds of Pic Nic as caretakers of the estate. Some of the younger Koehler children were born in the caretaker's house on the grounds. Mrs. Elizabeth Koehler died in 1912, but her daughter, "Miss Charlotte", a sister and two brothers lived on the estate until the sale took place in the 1950's.[1]

Some reports say the Schenleys lived in Pittsburgh from 1854 to 1863. This is not true. They returned to Pittsburgh in 1854 and sailed back to England in 1858.

The older Koehler children knew the lively Schenley girls and the dashing boys in the great splendid days of parties and balls. Miss Koehler remembered listening to her mother as she relived watching the Schenley girls dressed in the Civil War style of voluminous hoop skirts. Those memories of beautiful dancers whirling around in the ballroom at Pic Nic would live on forever with one who had been there. In the 1930s, "there was still much around the old mansion to remind 'Old Timers' of the day when Mary Schenley came back from abroad with her English husband and English servants to visit her home city. The ballroom's elaborate chandeliers and polished floors gleam just as they did in the days when dashing young Civil War officers and girls in voluminous dresses trod stately measures to the music of string ensembles."[2] The Johann Strauss waltzes were very big at that time.

Mary Schenley in her later years (courtesy of the Photographic Archives of Historical Society of Western Pennsylvania)

Lily Schenley (From the collection of Historic Locust Grove, Inc.)

Jane Schenley (From the collection of Historic Locust Grove, Inc.)

After the Schenleys' return to England in 1858, the twenty-eight room grand old mansion with English gardens stood alone on 105 acres and very proudly from the high hill. Sitting on the wraparound porch, one could view the Allegheny, Ohio, and Monongahela Rivers and Pittsburgh below.

The front passageway at Pic Nic entered into a wide hall where broad curving stairs mounted to the lofty second floor, creating a grand sweeping scope and walls 20 feet high or higher.

The ballroom was breathtaking and famous for its white Italian carved mantel and the elaborate huge dripping crystal and bronze chandelier weighing two tons which William had brought from Italy. Ornate ceilings, mirrored walls, hand carved Grecian columns with shining white oak floor made a palatial setting for a majestic evening of dancing. All the carving and plaster work was done by Swiss artisans. Many, years later, it was discovered that panels which appeared to be windows, but with a little pressure slid into the wall to reveal doors that led to bedrooms off the ballroom.[3] Precious oil paintings in their gold frames, paintings of Mary Schenley and the dashing captain who served at Waterloo were also to be seen.

The mansion was left with caretakers for over sixty years, but before the wrecking ball took over in 1944, the "Oval Room" and "Ballroom" were moved to the University of Pittsburgh where the ballroom is used for a meeting room.

In 1858, the Schenleys left that life in Pittsburgh and very soon after their return to England, Captain Schenley was elected to Parliament after an active campaign from a borough so small that the vote was 123 to 116. That brought another scandal, and he was tossed out on charges of buying votes. After all the investigating a committee reported that he had paid as high as $375 for votes and $125 not to vote: and that Mrs. Schenley bought one vote by leaving a pound note on the table of the public house.[4] William Croghan was correct when he speculated about his son-in-law's ways concerning money. Upon William's first visit to Mary in England, he had written to his sister-in-law that he was convinced Schenley was in debt and gambled. Mary must have known her husband's habits, too well, but why was that last vote left on the table of the public house? "She stood by her aged husband to the end."[5]

Winter of 1862 brought tragedy to the Schenleys. Their son William died at age 16. The following letter was sent to Mary's Aunt Elizabeth, written by Captain Schenley:

London 4th, Jany 1862

Dearest Mrs. Denny

The outside of this letter will convey to you the affliction which it has pleased God to visit us with the dear darling, "Idol" of us all, (Alas, so great—our love) left us to mourn him at half-past nine last night after 10 months of such suffering as I doubt fewer [?] been equaled by mortal, and the last six days so excruciating that it almost cracked on my heart strings, yet not a tear or complaint crossed his patient lips— "Mama & Papa are beside me and they will take care I recover soon"

Mary is in as sad state of stupor. We are all desolate—and now fear with [illegible] for her health. I can write only on this subject. Is it not a mystery that one so young, so innocent, so loved, then to have been doomed to unheard of suffering? It is difficult not to murmur. Will you write to Mary? She loves you better than anything now left on earth. Would that you were near her.

Yours affectionately

Edward W.H. Schenley.[6]

Mr. Schenley talked of crossing the Atlantic as soon as he could settle his family "as he thinks he ought to be there during this frightful struggle of our country. If he could do any good by going to Pittsburg, I should be most willing to let him go—but he is too old to take active part in the war & therefore his good wishes and donations & prayers for the success of the Union cause, which is affectual from this side of the water. The baby very flourishing on his eight teeth cut. Mary Schenley."[7] Mary's three oldest girls were all taller than she was and ages of children ranged from infant to nineteen years.

December 1862, the Schenleys were in Paris for Agnes's health. Lily and Jane brought their horses out with them and rode every day with their father. They planned to return to the U.S. in April..."but, I must say, I do not much like the idea of taking a helpless flock of girls to a country so disturbed and almost certainly of being again, laid up with asthma, having been exempt from it ever since I left Pitt. You may suppose how I dread a return to my enemy. But, with the exchange so high, we get very little of our income and would find it difficult to remain abroad. . . .'[8]

CHAPTER SIX

Mary Wins Over U.S. Courts

On January 31, 1878 Wyndham Harrington Schenley died at the age of seventy-nine. A year later, fifty-one year old widowed Mary claimed her fortune. It was not until January, 1881, that Mary finally won the legal battle she had started thirty years earlier when she became twenty-one.

In 1880, after a sharp court clash, she won a ruling that the 1842 law was unconstitutional, and she was entitled to her full estate. Morrison Foster (1823–1904), brother of Stephen Foster, wrote the following:

> The final Court Decision was handed down by the U.S. Supreme Court in 1881, giving Mary Croghan Schenley full control of her estate in Pennsylvania.
>
> On the 16th, of March, 1842, the Legislature of Pennsylvania sought to disinherit this daughter of the Revolution by passing an act which is a disgraceful blot on the statute book, for the purpose of depriving her of the control of the property left to her mother by the will of her grandfather, General James O'Hara. The Supreme Court of the state decided:
>
> 1. That the said Act of March 16, 1842, pretending to extend the trust under James O'Hara's will, was void.
>
> 2. That the trust had ceased; that the use had been executed, the entailment barred, and the title of Mrs. Schenley absolute—but to remove all doubts that the surviving trustee, James O'Hara, should convey to her.
>
> 3. That her title to lands under the sixth clause of the will was an unlimited fee simple.

Besides affirming the above propositions the Supreme Court in the opinion filed said: "As to the Act of March 16, 1842, we are clearly of the opinion that it was *unconstitutional*, nor do we see anything on the face of the bill or answer which would justify us in inferring that at the date of that act or indeed since, the appelle, (Mrs. Schenley), had become a British subject." She is entitled to the benefit of all the provisions of the Bill of Rights in protection of her property.

Decree affirmed. It had been contended in argument that having married an alien and lived with him in Great Britain the greater part of her life she became "ipso facto", an alien. Chief Justice Sharwood said "She could become an alien, only by naturalization."

Mrs. Schenley has never closed her house in Pittsburgh, but often has returned to it, and is as accessible to legal process or notice as any other citizen of Allegheny County. All her property in Allegheny County was devised to her in the will of her grandfather General O'Hara. Up to 1881 the title was tied up in a trust and the power to convey obscured. The trustee made his deed in January, 1881.

Since then, Mrs. Schenley generously made a present to the City of Pittsburgh of the Redoubt built by Colonel Boquet in 1764, adjoining Fort Pitt, together with the lot of ground 80 by 120 feet. The Councils of the City never accepted the gift.

Since February, 1881, Mrs. Schenley has sold a large amount of property in the Cities of Pittsburgh and Allegheny to different purchasers, about $150,000 worth, but the greater part of her property had been leased for terms running from 10 to 21 years at the time her title was cleared, and could not be sold.

It has been proposed to pass an act which would "deprive aliens of the right to hold real and personal estate, and to provide for the disposition of lands now owned by non-residents, aliens."

This proposed Act is a drive at Mrs. Schenley and her children. Such legislation, if it were constitutional, which it would not be, would also disenherit the grandchildren of General Grant, whose daughter married an Englishman and now lives in England.

Captain Schenley died in 1878. Since then Mrs. Schenley has passed part of her time in England and a part at Calais in France. She is now an old lady, and as a lady is a credit to her patriotic ancestry.[1]

Mary continued to live very comfortably in England with her grown children. At one time, $1,000 a day was being sent out of Pittsburgh to England. This pained Pittsburghers greatly.[2]

A Schenley granddaughter, Alberta McLean, wrote that after grandpapa died, the winter exodus from Princes Gate would end in Cannes where her grand-mama bought a lovely villa, Mont Fleur. Her grandmama and Aunt Hermione, their respective maids and a courier went off to Paris. A dutiful son-in-law would see them off and Mary Schenley's doctor went along in case she was upset by the Channel Crossing or her heart got tired. They spent a week in the best hotels in Paris which would give time for dressmakers (who went along) and the rest of the retinue with plate, linen, china etc. to shift from London and having everything ready at Cannes for Mary's arrival.[3] Mr. Carnegie often visited Mrs. Schenley at the beautiful villa in France.

The granddaughter never knew why her grandmama never went back to Pittsburgh, in later days. She was ardently American and Pittsburghian. During the Spanish-American war, [April 1898–February 1899] when the papers arrived there was a hushed silence while her grand-mama read the news. At American victories she gave her family cheques.

> It was always rather an ordeal staying at Princes Gate—the ceremony, silence and solemnity and there were so many of the family and grandmama never mixed them up—but I do wish I had registered more impressions of the friends and relation one saw there. One I do remember vividly, Andrew Carnegie; somehow he and Grandmama matched. Two tiny, vivid and vital people in the vast, dark plushy rooms—fresh complexioned, snowy haired (Grannie's was a wig). His eyes were so animated and his hands! He used them so marvelously describing a swimming pool that he was building up at Skibo Castle that one fairly saw it by the sweeping gestures in the air as he outlined....
>
> Thick, deep carpets, shining parquet, black and white marble, chilly white marble statues, dark heavy velvet hangings—all so overpowering, silent ceremonious and plushy all centering round

a tiny, Dresden-china complexioned little old lady. But WHAT concentration, grasp of a situation, power and knowledge she had—I think she was lonely for Grandpapa and the days of Picnic again....

Sometimes she would reminisce of her childhood days, romping into the country, riding through the woods at Pic-nic—"Dancing at pic-nic."

But, there was so much reticence, so much atmosphere of a person-apart about Grandmama, so tiny, so quiet, so dignified dressed in dark velvet or silk, priceless lace at neck, two big diamond drop ear-rings which flashed soft fire when the light caught them in the dimness with which I always associated Princes Gate.

"Cousin Julia" (Jesup) who had a marvelous voice and trained to Grand Opera form, [I] can remember her singing and it filled the big drawing room. George Rogers Clark comes into the ancient picture somewhere.

What history went with her [Mary]! If only she would have told it to someone to write down but that reticence which was so typical of her prevented her from making conversation of herself, her life and experiences. What a lot of absorbing interest she could have told us!

Photos, papers, news of Pittsburgh were such a joy to her. She was always thinking what she could do for her home-town—possibly it was a heart supposed to have been a bit strained, or fear of the Atlantic being rough. Was it that she did not like the idea of going back to old familiar places with the blanks of loved faces, no longer there?[4]

One day Mary didn't come down for lunch. The maid had heard a quiet voice say, "Oh, I don't feel well," and as they turned to her, Grandmama gently went limp on her pillows. After Mary Schenley's death, her home, No. 14 Prince Gate, was sold to Pierpont Morgan, who already had No. 13. The two were thrown into one to make more room for his art treasures. Then it became the U.S. Embassy.[5]

Mary died November 4, 1903 and funeral services were in All Saints Church where she had attended for many years. She was buried in Ennismore Gardens, London. When the news of her death was heard in

Pittsburgh, Mayor Hays and joint sessions of Pittsburgh Councils met to honor Mrs. Schenley. Never before in the history of Pittsburgh had such an honor been accorded a woman.[6] The first action was to erect a permanent memorial for Mrs. Schenley in Schenley Park which she had disapproved during her lifetime.

The opening line in her will begins as follows:

> I, Mary E. Schenley, nee Croghan, a citizen of the United States of America, of Pittsburgh, Pennsylvania. ...[7]

Her American executors were Andrew Carnegie, and John W. Herron, of Pittsburgh; and Denny Brereton, of Yonkers, New York. The first name for English executors was her only surviving son, George Alfred Courtenay Schenley.

Prominent men in Pittsburgh who came in contact with Mrs. Schenley, in regards to her vast estate, said that "She was a remarkably shrewd woman and one of the best business women they had ever met."[8]

The youngest daughter, Octavia Hermione Courtney Schenley, was her mother's constant companion until her death in 1903. She must have felt a great loss, but on the other hand, she was free to travel, and that she did. She was still single and visited Pittsburgh in 1905.

Hermione had recently been staying in Scotland with the Andrew Carnegies at Skibo and before that, she was with the Alfreds on their yacht in Holland. Arrangements had been made for visiting New York and Pittsburgh and going on The Cunard Line. She took Edward Harbard [Lily Schenley Harbard's son] and he seemed very pleased to be going, but both were expecting the weather to be very cold in November.[9]

In 1906, Hermione attended the Louisiana Purchase Celebration and on that trip she visited Clark relatives. Upon her return to England, at forty-eight years old, she married Edward Downes Law, Baron of Ellenborough, Commander of the British Navy.[10]

In 1910, George VII took the throne of Great Britain and Hermione was included in all the pageantry in a royal way since she now had the title of "Lady Ellenborough." She wrote to Julia Jesup from Wimblesham Court, Surrey, that she and her husband would be leaving London on the 8th of May, where they would be gay and busy. After the Court on the 24th, they go to the Coronation on June 22. Husband Edward had all his grandfather's robes, but there was only the crimson velvet kirtle left of the 1st Lady

Ellenborough so, she would have to get a new one which would be white and gold brocade.

Hermione continued to write about taking her seat amongst the Pureses in the House of Lords. She would always attend when Ellenborough was to make a speech. Again, Baroness Ellenborough [Hermione} arrived in New York in 1914 and had a long visit in Washington D.C. Upon her arrival in Florida in January, she received the news of the death of her oldest sister, Lily Schenley Harbord [1843–1914]. Lily's son Edward Harbord, who had accompanied Hermione on her trip to America in 1905, was then locked up in a fortress in Germany, a prisoner of war. The "Lusitania" had sunk since their arrival in U.S. and they had to return on another vessel which was a sad homecoming, especially as poor Edward Harbord was shut up in a fortress and would not see his mother again.

Lady Ellenborough arrived in Pittsburgh once more on April 28, 1926 accompanied by her nieces, Mrs. Finch [Jane's daughter] and Mrs. Forster [Lily's daughter]. They were there to observe the founding of the Great Carnegie buildings which graced the entrance of Schenley Park.

1931—The MARY SCHENLEY COLLECTION of Victorian furniture and furnishings were for sale at Kaufmann's, eleventh floor, under direction of The Galleries. In Mary Schenley's will, she directed "Picnic House" would always be kept open as an American hospice for her heirs across the sea. The heirs never occupied the house for long. The trustee, with consent of the heirs, decided to dispose of the furnishings. It was said the collection was one of the choicest examples of Victorianism existing; a letter-perfect picture of 19th century taste.

Among the heirlooms were some which Mary Schenley's renowned grandfather, General O'Hara, had received from his friend, George Washington, and in turn, had been left for his grandchild. There were things which were used when the General O'Hara had entertained his distinguished guest, Louis Philippe, heir to the throne of France. Among other items were things presented to Mary by Henry Clay. Some valuables were presented by the Schenley Estate to Carnegie Institute.[12]

Three hundred acres of hills, valleys and streams were given to Pittsburgh for a park. No city was in need of a park more than Pittsburgh and with these acres, no one had finer lands than this, which was her first gift of land given in 1889. She also gave part of Riverside Park and the sites for West Penn Hospital, the Blind Home and the Newsboy's Home.

All this generosity bestowed on Mary Croghan Schenley's beloved Pittsburgh, described herself in her will as "of Pittsburgh, but now residing in England." All this began with Mary's Irish grandfather, James O'Hara. He came to the new land before the Revolutionary War to seek his fortune. He started in the salt business and was so successful he built Pittsburgh's first brewery. He became a director of the city's first bank and he started the first manufacturing center with his glass factory. But, most of all, he bought land.

James O'Hara was a quartermaster general in the Revolutionary War when he helped to put George Washington in as President by being one of the country's first Presidential Electors.

At the Oakland entrance to Schenley Park stands Carnegie Institute and farther on is Phipps Conservatory, the second largest building (two and half acres) under glass in the country. Among its flower displays is an orchid collection that is one of the world's most famous.

During World War II in England 1941, Hermione was still living at age eighty-eight in Sunningdale, Surrey, England. She was happy to report to her nieces and cousins, she had not been bombed yet. About that time, Mrs. Harmar D. Denny, Jr. from Pittsburgh said that during a visit, Lady Ellenborough greeted her with the words: "Tell me about my dear Pittsburgh and my dear cousins there."[13]

In 1949, Mary Croghan Schenley's granddaughter, Alberta McLean, living in New Zealand, wrote about the event that rocked two countries in her "Memories". Her grandmama and cousin Lucy Jesup were at school together in New York. The following is written by the Schenley granddaughter.

> Grandpapa went to the wars very young. He was an Ensign at Waterloo aged 15! Now, as to Grandpapa's parents; there was a little place near Barnet (ten miles out into the country) called Schenley. I once found myself motoring through it and looked at the little white house, more like one of the small manor-farms than anything else. I always wanted to hunt around Ireland to look for Croghan's Gap where family report says the original Croghans came from—the O'Haras coming from Charlesville Forest.

> Great grandmama was a beauty. (I think her maiden name was Fitzgerald), her husband just a small country squire. Yet in days when commissions in the army were acquired by purchase and the price rose according to the smartness of the raiment, there was grandfather in the crack regiment of the day—The Rifle Brigade.

I believe he was wounded in the leg at Waterloo refused to have it off saying his legs would both live and die with him.

When the grandparents went to live in England, they entertained all the top-notchers, stayed at Blenheim with the old Marlborough,—there are photos of them in the album and one of a small, pudgy boy in sort of Fauntleroy suit labeled Lord R. Spencer Churchill, Winston's grandfather. Well-Well, great grand-mama was very, very lovely and if the whispered conjectures amongst my mother and her sisters and brother had any founda-tion, she must have touched the heart of someone in the days when royals could not run away with commoners wives!!! At any rate, it seemed as if somebody did a powerful amount of "fairy-godmother or father" to grandpapa. If you are interested in her-aldry it might be interesting to look up the crest of the old carved ivory seal that I sent to the Historical Society of Western Pennsylvania. I meant to try and do this once, should have looked up the house of Hanover in Almanach de Gotha.

This is just the gossip overheard amongst the aunts and uncles and sometime of later years discussed amongst us grandchildren.[14]

This written piece by the granddaughter seems to be a summary of the whole beginning of this saga. If the fortune hunter had not come to town, Mary Croghan would most likely, have been mentioned only in a family chart. This elopement affected so many lives in more ways than one and Mary, like her two Irish grandfathers, Major William Croghan and General James O'Hara, would always be remembered for their place in history.

Today, at the entrance of Schenley Park in Pittsburgh is "The Mary E. Schenley Memorial." It represents "A Song to Nature" and stands in the center of the former Bellefield Bridge.

The irony of all this is: it all began with Mary's Irish grandfather, James O'Hara. "None of this historic importance bears his name. Only the name of the Englishman, Schenley, who came to America and stole his grand-daughter away, is remembered."[15]

Through the years, Pittsburgh residents have expressed the possibility of a movie being produced of this remarkable romance that rocked two nations. One newspaper writer suggested, "In THE TRUE-LIFE romantic drama that ensued, Picnic House was a seldom-seen backdrop. If the

Croghan-Schenley romance was made into a movie, the mansion probably would be featured briefly in the first reel, or portrayed in flashbacks as an ill-focused memory—like the wintry boyhood home of Citizen Kane."

PART IV

Widow of Lafayette Square

CHAPTER ONE

At Home in Kentucky

One week after the Civil War ended in April 1865, Mary Jesup Blair witnessed a very poignant scene which took place in front of the Blair House, Lafayette Square, Washington D.C. Across the street, President Lincoln was standing on the portico of the White House. By the President's side, stood a relative of Mary's from Kentucky, Major Robert Anderson of Fort Sumter fame. Other Union officers were also present and all were viewing Lee's surrendered Confederate Army, now prisoners, passing on their way to stockades. One of those prisoners was Meriwether Lewis Clark, son of General William Clark of the Expedition fame. Both Mary and Robert Anderson were related to Meriwether.[1]

Anderson, Clark and Lincoln had served together in the 1832 Black Hawk War. Family letters reveal an incident which took place between Clark and Lincoln one day when marching out of Jefferson Barracks during the Black Hawk War. Clark was riding near the rear when he noticed by his side on a small horse a long-legged, dark-skinned soldier, with black hair hanging in clusters around his neck, a volunteer private. "Admiringly the private gazed at Clark's fine new uniform and splendidly accoutered horse." Young Clark spoke to the poor soldier, who asked him many questions such as, "Are you the son of Governor Clark of the Lewis and Clark expedition and related to all those great people?" Clark answered with a "yes," but before moving on, he felt an interest in this lank, long-haired soldier and turning again asked him where he was from, and to what troop he belonged. He said, "I am an Illinois Volunteer." Clark then asked him his name. "My name is Abraham Lincoln, and I have not a relation in the world." Clark's heart had gone out to the soldier in pity and sympathy.[2]

The next day after the defeated soldiers had marched in front of the White House, President Lincoln was killed, which caused Meriwether Lewis Clark much sorrow.

Mary Serena Jesup Blair was viewing the march when suddenly Meriwether passed by and managed to hand her something and exchanged a few words. Evidently, it was a request to contact their Uncle George Hancock in Kentucky, who was uncle to both Mary Jesup Blair and Meriwether Lewis Clark. Mary immediately wrote and received the following letter:

Louisville, Ky. May 3, 1865

Dear Mary

I have this moment received your kind letter of 30th Apr. and thank you for the kind and prompt action in behalf of Lewis Clark and I hope your efforts succeed in accomplishing his release. If it can be done at all, I know it can be, by the influences that you can bring to bear.

You had better retain the articles he handed to you, until you find whether your efforts succeed on which event they can be returned to him, so dispose of as he may direct.

Inform me the cost expended for him, & amount of money furnished, and I will send you a check on N.Y. for it.

I write with difficulty, owing to partial paralysis of my right arm; which will excuse for a short letter, when there is *so much* that I would like to say to you.

Permit me affectionate remembrances to all with you, and believe me as always truly yours. Geo. Hancock.

—Mrs. Mary J. Blair Washington City.[3]

Meriwether Lewis Clark was released immediately and Mary would have taken him to her home on Lafayette Square.

Mary Serena was thirty-nine years old at the time this event took place. How did she gain so much influence with the Washington D.C. powers?

Influence and power start with old Virginia ancestors and Mary Serena was blessed with famous nation-building ancestors. Her father had held a position for forty-two years of service as quartermaster. The General's office was in the White House and he had witnessed the arrival and departure of ten or eleven Presidents from James Monroe, beginning in 1817 to the end of James Buchanan's term in 1860.[4]

Mary married into the very powerful and influential family of Francis Preston Blair, who was builder of the famous Blair House in Washington D.C. and he was the only newspaper editor in Washington for several years. Blair was from Kentucky and had been a schoolmate of the Croghan boys.[4]

Blair's editorials could sway and influence the outcome of the happenings in government and close family connections existed between the Blair and Jackson households. Blair's daughter Elizabeth Blair, who later married General Robert E. Lee's nephew, Phillip Lee, lived in the White House much of the time during Jackson's term. President Jackson wanted her there since the conditions and atmosphere in the White House were beneficial to Elizabeth's fragile health problems in contrast to the dampness and draftiness at the Blair House across the street.[5]

Mary was a newborn baby in 1825 when Grandmama Croghan came to live with the Jesups in Washington. The greater part of the next thirteen years, Grandmama Croghan and Mary Serena Jesup were living in the same home, whether in Washington D.C. or in Kentucky.

May 20, 1836, Thomas Jesup, brevet of Major-General, assumed command of the army in the Creek nation, and in December he took command of the Seminole War in Florida.[6] This assignment was expected to be completed in a short period of time. The war started when the U.S. Army was sent on a second campaign to Florida swamps for removing the Seminoles to Indian Territory out West. But, for four years, Ann and children lived at Locust Grove in Kentucky for the duration of the War.

Before the Jesups left Washington for Kentucky, William Croghan of Pittsburgh was suggesting and advising Ann as to the best means of traveling to Kentucky. He was recommending she visit Pittsburgh instead of Wheeling because the river was then flooded and no doubt would be high and navigation would be interrupted by ice. However, she could get the choice of staterooms and berths. William assured his sister that the Pennsylvania Road and the Nation would be worse and accommodations were not good. Pittsburgh had a line of Packet boats that left there daily for Louisville. William urged his sister and party to stop and stay with him where he could make her and her party very comfortable until one of the favorite boats takes its departure. "I can not tell how happy I would be to see you all here".[7]

Grandma Croghan and Uncle John would be waiting with open arms for the arrival of Ann and her brood in Kentucky. Uncle George Croghan was at Locust Grove when Ann arrived with children in January 1836. Jesup was in the swamps and expectations to capture "Osceola" had failed even though he surrounded himself with competent soldiers, such as Zachary Taylor.[8]

In the meantime, Ann Jesup and the children's stay at Locust Grove would last for four years. Uncle John and Grandmama Croghan became even more attached to Ann's children and Locust Grove became home for the children.

Doctor John wrote to Jesup, "Yesterday, Judge Rowan, Judge Bibb, Judge Pirtle, Prentice, the editor of the Journal, and one or two more dined with us, and Rowan offered a toast complimentary to you ... Billy boy and the Big Man walk a good deal with me over the farm and occasionally we pay old *Uncle Jim* as he is called, a visit at the Mill. He feels highly honored at our visits. Although it is some time before our breakfast, the *Big Man* is seated to my left, talking away and ready to take with me our accustomed cup of coffee."[9]

An excellent musician rode out every Monday evening to instruct the "little lady" [Lucy Ann Jesup], Mary, & Jane. Uncle John was finding it difficult to procure a governess, and at last made known his wishes on this subject in one of the daily papers. He thought the nieces were studious for persons of their age and had improved a good deal. Judge Rowan of Federal Hill, Bardstown, Kentucky, *My Old Kentucky Home* fame, was assisting and providing necessary information respecting the Female Academy near Bardstown and Guthrie, likewise respecting the Cincinnati schools.[10]

Charles Jesup was born March 1835 and was not yet a year old when Ann brought the children to Locust Grove, so the first four years of his life, Locust Grove was his home. No wonder Charles was "Uncle John's" favorite of all his nieces and nephews. He especially, enjoyed Charles when he performed his whistling act for his uncle.[11] In later years, Uncle John bought a Jew's harp and sent it to Charles for his performances.[12]

Once again, sounds of laughter, music and dancing rang in the old ballroom. They too, would have joined Clark cousins at different homes to form dancing sets.

At Locust Grove, the Jesup children heard stories of Grandmama Croghan's childhood in Virginia and their move to Kentucky on the Ohio River in a flatboat. When the Jesup sisters were much older, they wished they had been more attentive while these stories were being told, but in youthful years, attention placed on family history is minimal. Lutie wrote to Mary that she remembered Grandmama talking about visiting her Grandmother "Bird's' house and the girls could recall a remarkable amount of family history. They came to know the Clark relatives very well where each belonged in line of descent and to which family.[13]

May 1837, Grandmama Croghan, Ann and children were visiting in Louisville. Their purpose was to attend the nuptials of Martha Pearce, who was married on the 17th to Mr. Stannard of Virginia. John was having a small room adjoining the house added as a chamber for Lucy and Ann had suggested they stay in Louisville until the plasterers and painters had finished.[14]

George Croghan had been with the family at Locust Grove for a while, but was at that time in Nashville where he planned to visit President Jackson at the "Hermitage" before his departure. John wrote that George behaves, and had for a long time, as they would wish. He had been ordered to muster into service and discharge the troops of Kentucky and Alabama.[15] This could have been when the first short separation between George and Serena took place.

In the spring of 1838, Mary Serena was thirteen years old and witnessed a sad time at Locust Grove. She was with her grandmother in her last days before she died on 4th of April and was at her graveside when "Grandmama" was buried beside her loved ones in the old family graveyard.

John and Ann received a letter from George.

Washington 17th, April 1838

Dear Dr.

I have been out today for this first time unless to church, since the receipt of the distressing news of the death of best of Mothers. I feel more than you all her loss, for I have in addition to our common griefs, the agonizing reflection that I observe of all her children have by my repeated misconduct caused her anguish & distress. She is gone & I can not offer her this [illegible] assurance that my ways are changed; but in affectionate rememberance of her every prayer & wish I will henceforth strive to act as I know she would desire were she on earth.

G. Croghan.[16]

That wasn't an easy time for Ann Croghan Jesup. The death of Lucy coincided with Jesup's embarrassing situation when he made the decision to reject the traditional rules of warfare in Georgia.

General Jackson had sent Brevet Major General Thomas Jesup to the Seminole country, to replace Major General Winfield Scott who had failed to remove the Indians. One of the subordinates with Jesup was Zachary Taylor.

Jesup seemed to be more interested in the "slave catcher" business while in the Seminole country. Runaway slaves and their descendants had been accepted by the Southern Indian tribes. Slave trader's raids on the "Negro Seminole" became common. By mid-1837, General Jesup had returned approximately 100 Black Seminoles "to their righful owners."[17]

Hostilities became once more a composite of guerrilla attacks, scouting expeditions and slave-hunting raids. Jesup cast aside procedures of war and "awarded seized 'property' to the looting detachment and even gave a reward for Negroes brought in alive by the Indian allies. The commanding general himself acquired an unenviable record by making one of the largest hauls of the war."[18]

In the spring of 1837, Jesup had arranged negotions with the Seminoles. An agreement was met and "in March Jesup was so confident of a settlement that he reported to his superiors, 'The war is over.'"[19]

This statement was wrong, the Indians just faded back into the swamps. Jesup threatened to import bloodhounds to flush out the enemy. By this time, "Jesup had lost credibility, and nothing he promised would convince the Seminoles to emigrate willingly. Both Congress and Jackson wasted no time in heaping abuse upon him."[20]

Jesup again opened negotiations by offering the Seminoles a permanent home in South Florida to the Indians. President Van Buren sent word that this couldn't be done. Jesup knew peace terms would only mean defeat for himself and neither was he willing to allow Osceola to escape so he called a conference with Oscela for truce. It was too tempting, now that he had the enemy in his presence. He quietly had his soldiers to circle the Indians and they were captured.[21]

He bagged the whole group, "513 Indians and 165 Negroes," and solved that hold-up. Jesup was replaced by Zachary Taylor. Seminole Chief Osceola was sent to Fort Moultrie prison (S.C.) where he died of throat infection in January 1838. The chief was buried there with full military honors but minus his head.[22]

The war chief was buried as a hero. In the meantime, Jesup suffered ridicule and derision as the "personification of treachery throughout most of the civilized world."[23]

Jesup must have had visions of delivering Osceola to Washington D.C., triumphantly as General George Rogers Clark had delivered British Officer Hamilton to Virginia during the winning of the Northwest. There was one outstanding contrast between the honorable capture made by General Clark and the capture through tricks and treachery used by Jesup.

In the spring of 1838, he had to turn active command over to Colonel Zachary Taylor. Some senators questioned Jesup's conduct of the Seminole War and provoked an inquiry. The chairman of the Senate Committee, Thomas H. Benton of Missouri, intervened and stopped the inquiry. Congress ridiculed him and Jesup never regained the credibility he once had.[24]

CHAPTER TWO

Beaus and Belles of Washington

By 1840, the Jesups were back in Washington City. In the meantime, Ann Jesup's brother William Croghan had become very wealthy and wanted to send Ann horses for her own use since it was the custom for ladies to have their own conveyance and, evidently, Ann had none for her personal use. A coach was commonly used for getting about town.[1] Dolly Madison could be seen quite frequently going about the streets of Washington in her carriage.[2]

Dolly Madison could also be seen seated on the front steps of the Dolly Madison House laughing and talking and surrounded by young girls.[3] The Jesup girls could very well have been present at Dolly's house on those summer afternoons.

The popular Jesup sisters whose father was one of the powers of Washington City were living the best years of their lives in 1840. They partied and danced at the White House until wee hours in the morning. When celebrating Harrison's inauguration, on the 4th of March, 1841, Lucy Ann was eighteen and the following year she would be attending school in New York. Mary was sixteen and Jane was fourteen. There's no doubt, the gowns for all the grand balls required much time spent with the dressmakers in choosing and fitting the right gown for each occasion.

Uncle William Croghan of Pittsburgh was a frequent guest of Ann and family and was always in Washington City for inaugural celebrations. Every four years a new elected president would be sworn in and Clark relatives from far and around would descend upon D.C. for inaugural celebrations and "drop in" at Cousin Mary Jesup Blair's home on Lafayette Square. It would be open house for relatives and friends at any time during the day for food, and her home was the best location for viewing the new elected president as he rode to the White House. This custom lasted until 1933 when the Lafayette Square celebrations ended after Violet Blair Janin's death.

When Harrison came into office, the new railroad steam cars had just come into use and from this time forward, the railroad cars and the steamboat made travel much faster and more comfortable. In February 1841, the newly elected President William Harrison boarded the Baltimore & Ohio railroad car in Frederick, Maryland, to Baltimore, arriving at the capital.

President Harrison's widowed daughter-in-law, Jane Findlay, would be hostess of the White House since the president's wife stayed in Ohio. "All women of means" had come elaborately equipped and Mrs. Findlay had brought her coach, horses, and coachman.[4]

The following describes Harrison's grand entrance into Washington:

> On March 4, astride a magnificent white charger, he rode to his inauguration at the Capitol, amid marching militias, floats laden with log cabins, maidens in white carrying garlands and signs with Harrison slogans, and a crowd estimated to be second only to that which had attended the inauguration of Washington.[5]

Winter of 1844–1845, James Blair, son of the famous Francis Preston Blair, was in Washington City for the season. This young man must have been one who caused heads to turn and take notice when he entered a room where others were gathered.

James Lawrence Blair was appointed midshipman in the United States Navy in 1836 and accompanied Commodore Charles Wilkes on his Antarctic voyage in 1838–1842. He then was a member of an exploring expedition on the Pacific coast and compiled *Notices of the Harbor at the mouth of the Columbia River*, which was published in 1846.[6]

In 1844, James Blair had been told by a young lady of Lexington, Kentucky that she would not become his wife as planned. The two families of James and the young lady had played an important part in breaking up these wedding plans. No matter how many angry protests James wrote to his family, especially his mother, and letters of accusations going back and forth, James threatened to leave the navy and come home to marry the young lady. His mother instructed his sister Lizzie to let James know, "that none of my sons shall bully (& the Lioness spoke) me into doing or approving of a wrong & if he is not man or gentleman enough to follow faithfully his honorable profession the sooner he gets to the [indecipherable] the better."[7] All the fighting ceased when the girl's parents refused to let their daughter marry.

James very quickly recovered while in Buenos Aires "riding 20–30 miles a day and dancing every evening in the home of some native don, where there was usually to be found a fine collection of native Spanish black-eyed beauties, who could waltz or dance with such enchanting grace that I was very nearly captivated."[8] However, he escaped this captivity and returned to Washington and to his surprise, he discovered in his long absence, the Jesup sisters had grown up into beautiful young ladies and he fell in love with the pretty one named Mary. This time, he was captivated for life by nineteen year old Mary Serena Eliza Jesup, who was his one true love.

In the spring of 1845, James had to leave Mary in Washington while he resided with President Andrew Jackson at the Hermitage in Nashville, Tennessee for several weeks while collecting and boxing President Jackson's papers in preparation for returning them to Washington, D.C. for depositing the papers in historical archives. His sister Elizabeth Blair from an early age had always been a great favorite of Old Hickory. She and her girl friend accompanied James for this visit with the Jackson family at the Hermitage. President Jackson would have been very disappointed had they not arrived.[9]

This was the summer (1845) the Jesups came through Louisville on their way to Mammoth Cave in western Kentucky for a stay of several weeks.[10] Mary Jesup, with her family, were waiting at the Mammoth Cave for one special guest, James Blair, who was expected to arrive from Nashville.

In this romantic setting at Mammoth Cave, in September 1845, James Blair finally arrived. He was in love and one can imagine the romantic times in the ballroom, walking on the veranda and the colonnades in front of the large hotel, which extended the whole length of the long building. Probably a beautiful Kentucky harvest moon was shining when James asked Mary to be his wife, since this was where they became engaged.

James informed his father, Francis P. Blair of his engagement to Mary Jesup Blair and then, Ann Croghan Jesup received a letter from James's father in which he informs "My Dearest Madame" that he had just received a letter from James who gave him great happiness and hope to know that their two families would be intimately related since they had a long unbroken friendship. He had a great affection for Ann's Croghan brothers, since they were schoolmates long ago. He wrote, "You remember that when you came to take leave of us, I asked you to give Mary to me." He assured Ann that the Blair family would have great affection for Mary and James would

make a good husband and that this father would take much delight in cher-
ishing her as if she were their own offspring.[11]

The frontier bloodlines would be united and this wedding served as a
perfect opportunity for assembling blood kin and friends from Kentucky
such as Henry Clay. Calhoun, Webster, Buchanan, and President Polk were
only a few of the many political friends who were in attendance.[12]

> I certify that on the fourteenth day of January: eighteen hundred
> and forty six, I united in marriage, according to the forms of the
> Church in the U. States.
>
> James Blair & Mary Serena Eliza Jesup at the house of her father,
>
> Major Gen. Jesup
> Smith Payne Rector of St. John's Parish Washington D.C.

Henry Clay of Kentucky had been a schoolmate of Francis Blair and the
Croghan boys. Blairs and Jesups were part of the well established first families of D.C. society.

Three months after the wedding, Dr. John Croghan received a letter in
April informing him of his beloved sister Ann Croghan Jesup's death, April
1846.[13] At that time, Dr. John lived alone at Locust Grove and the grief
must have been harder to bear. There were only three of Lucy Croghan's
children left: John, George and William, the three oldest brothers.

Ann's youngest child, Julia was six years old—the boys were eleven and
thirteen when they lost their mother. In September, Mary was expecting the
birth of a first child, so husband James escorted Lucy Ann, Janie, Julia and
Charley to visit their Uncle William at his large country estate Pic-Nic, in
Pittsburgh for a three week stay. James wrote to Mary reporting their travel on stage to a Tavern for a night, and from stage to Steamboat arriving in
Pittsburgh where they were met by Uncle William.[14]

They were very tired and went to bed early the first night. Six year old
Julia had not cried since they left home. James wrote, "she appears to keep
crying as a sort of *home entertainment*. She will not probably cry during
her absence and will therefore have a great to do in that way when she
returns home to make up for lost time."[15] The girls and Charley were very
happy. Uncle William was very attentive his conversation singularly fascinating and the girls would laugh themselves fat before their return if they
keep up in this same way for three weeks. The changes of place and air had
a happy effect upon them. A faint bloom returned to their unpleasant look-

ing marble-like paleness in the face. The exercise of the journey and mountain air had been helpful for all.

Grapes and peaches were there in great profusion and Charley and Julia were in ecstasies. Julia picked out large bunches for her Pa and "if he can't get them, he ought to have them", and means to send them to him by *Blair* and Charley.[16] That was the last time she referred to Mary's husband as *Blair* due to Mary's command.

The next morning, James had gone into town, seeking more information about a ship assignment for his next sea duty. Upon his return, riding slowly along to reach Pic-Nic, a young man dashed past in a buggy in a brisk trot. When they both turned off to together to rise the hill, the young man introduced himself as Wilkins, then dashed off to the house in a fast pace. He began to remember what Mary's Uncle William had told him that very morning that Wilkins had designs on Lucy Ann but, had a problem with dissipation. His father, at one time, denied him his own house. Uncle William considers it his duty to tell Lucy Ann these things. After tea the next day, James and Wilkins walked out together. James took it upon himself to tell Wilkins what he had heard and he did not deserve any consideration of expressing his attentions and was not worthy of Lucy. This talk to Wilkins was a big mistake and James learned a very important lesson from Mary and Lucy Ann, which he never forgot: "do not interfere with the Jesup girls' romances."

James returned to Mary to await the arrival of their child, Ann Jesup Blair, born December 10, 1846. But in February, he had returned to ship duty at Norfolk, Virginia and wrote to Mary that his thoughts were his darling wife and their sweet baby. "I can see you both in my imagination, joyous and happy in each other, but I am almost crazy to see you again, altho, it has been a few days since we parted." His mind and heart would be more at ease if Mary would walk out every fair day and gradually increase her exercise until she gained her strength for the baby and him. "Seek, my darling, the company of your sisters and all your friends in and out of the house. If I find that you will do this, I will then more patiently endure my absence from you."[17]

The Mexican War was beginning at this time and General Jesup was obliged to supply the Army in Mexico. James expected to return to D.C. before General Jesup's return.

Sometimes James thought he was the most impoverished husband in the world and yet, even though he tried, he had little consolation that he could rid himself of the many little debts that appeared to be always hanging over him. "But, if God spares my life, I'll endeavor to have you surrounded by every comfort that you may want and I'll be with you to see you and our little Anne enjoy them as we gaily and cheerfully move down the stream of time. I hope, my darling, that you will approach God's Altar with your sisters. You know I talk but little on the subject of Religion, but you know I love the truth and faith of Christ's Religion."[18]

James continued to write the most affectionate letters to Mary.

> Take care of yourself, My Pet. That our little Anne is so well and smart gives me great pleasure, but how I long just to see you both in your love and sweetness to each other and for me. I feel confident that our little Anne will be able to call upon her Papa before many months ... I can see her little mouth drown up and then a broad smile upon her whole face and her deep blue eyes. Oh! she is a perfect little *dew-drop*, pure, clear, fresh and lovely.[19]

Once again, James was with Commander Perry in Norfolk, Virginia. He said Perry was a man of great enterprise and daring but, no judgment—brown as a tiger and as ostentatious as a peacock. The letters were full of reports about the navy operations and keeping the ship in operational order. He spoke of hoping to succeed in obtaining the command, or the second in command, of one of the steamers lately authorized by Congress. The law specified that they must be commanded by Naval officers. If he succeeds, he will return to Mary every month.

While on ship, this young father received very sad news from Washington.

> Norfolk March 23rd
>
> My dearest Mary,
>
> I heard within the last two hours from Mrs. Frazier and I have seen the newspaper containing the announcement of the loss of our little Angel who has gone to a home of rest and of bliss. Poor little darling, she had suffered much in this world. She deserves a home of rest and relief from suffering. Mary, my heart is bursting with grief. God's will be done. I can now weep and mourn, my Pet, but my heart was too full at first. My prayer this moment is that God may

give you strength to bear the affliction. Our savior said, "let little children come unto me for of such are the kingdom of heaven."

Cheer up, my Mary, cheer up, be of good heart for our little darling is now in heaven in the loving of God and as with her own loving Grand Mother. She has to suffer no more. Therefore, my own Mary, be consoled, be meek under God's ... for all he does is for the best, altho, it may rend our hearts. I weep, now, Mary and my heart is somewhat relieved. I'll write again when more composed, my own sweet Mary. Take care of your health, take exercise and nourishing food for the sake and love of him who lives but to live in your heart. God protect, guard over and keep you in good health and spirits is the prayer of every moment of your absent devoted and affectionate husband who weeps too much, now, to write more. Kiss your sisters, tell them that I bear my affliction well, God knows I try to. God be with you and sustain you in this trying affliction.

Your loving husband J. Blair[20]

The letter to Mary was stained with tear drops.

U.S. Steamer Polk
Navy Yard March 24th, 1847

My dearest darling Mary, I have become more composed, not that I feel the less the visitation of God upon us, but my heart mourns within me and deep grief has settled silently upon me. Yet, I endeavor to be cheerful that I may not disturb the pleasure of those about me. I have looked at your likeness this morning, the first thing after I rose and then read the letter in the Bible you gave me and have just finished looking at the hair in my locket of our little angel who is now in heaven. Peace be with her darling little soul in her home of happiness. Mary, my heart weeps silently the unconscious tear is often on my cheek, even at this moment my full heart fills my lungs (?) so that I scarcely know what to write. Oh! that I could be with you to comfort you now with all the fondest most devoted love that God has given me power to feel, to draw you to my breast and hug thru rest—thy mourning spirit to end of time.

Good night, my pet, God be with you.

I have been several hours writing this—I know not what I (illegible) could have written all I have thought. I was composed when I commenced but, God knows I am not, now.

Good night, my Mary—God be with is fond prayer of your old man. JB[21]

Mary came to James in their time of sorrow. After her return to Washington, she received a letter dated March 31st, in which he wrote, "We started early the next morning after you left me and remained in and about Hampton Roads all day yesterday and last night. I was very anxious to know that you arrived home safely."[22]

James had signed on as an employee of the steamer *Washington* for a voyage to Europe, hoping to get an assignment to his advantage which was expectations of getting command in the next year or two. Conversations of his companions and associates at the dinner table made his letters to Mary much more interesting.

A trip to Germany, then to London for two days was made to see Mary's Uncle William who was visiting his daughter, Mary Croghan Schenley and family. James found Southampton very neat and quiet. Streets were beautifully paved and so clean, sidewalks were wide enough, smoothly flagged and kept perfectly clean. Very many of the houses had large bow windows extending from near the ceiling to the floor by which you can admit light through the whole front of the room or exclude all or as much as you choose from above or below. One could walk into the bow and see up and down the streets without raising the sash.

Wishing that he had insisted on bringing Mary along, he lamented that this could have been their honeymoon trip they never had. The sea sickness would have been worth the happiness to him to see her enjoying the beauties and comfort of this old world. His letters were filled with a driving desire to make the best life for his Mary. "I have determined before I left, my darling Mary, to work hard while yet young to place you beyond the possibility of want in case you should lose me. How happy I will be to see you enjoying every comfort and luxury of life and feel that I have given them from the fruit of my own exertion. Oh! Mary, there is downright pleasure in hoping that I will be able to do it."[23]

On July 1, 1847, James was in England and had seen Mary's Uncle William, Mr. and Mrs. Schenley, the little Schenleys. Mr. Schenley looked more like a man of only 38 or 40 and dressed neatly. Mary Croghan

Schenley was fully grown and would pass for 18 or 20 years of age. They had insisted on him taking his meals with them and had been very kind and attentive. Uncle William had been so attentive and insisted on going with him to point out the beauties to me. "I have passed two evenings in Hyde Park from 5 to 7 and the turn out of Fashionables is immense and the variety so great that it is almost beyond belief even as spectators."[24]

James took all this in stride as a perfect aristocratic monkey-show. The aristocracy in that country he wouldn't have believed, had he not seen it. Carriages with the Royal Coronet upon them and flying about town and the parks. The ladies were lying back in open carriages with their feet up on the front seat. Ladies were almost lying down in pairs in the carriages with music monkies and boy tigers dressed up in all colors of the rainbow. Any quantity of gold and silver ever was on front and rear of carriages and looked like light houses as a warning to get out of the way to prevent being run down by the carriages.

He had seen the Queen the evening before in the Park with a flaming red light horse several yards ahead of her, which was notice to everybody in carriages to stop and let her pass. Of all the monkey shows that he had ever witnessed, he thought this most extensive and perfect. James had begun to realize that he lived among the greatest people of the earth and was looking forward to seeing Mary in six or eight days.

After returning to Washington from Europe, James was back on duty in New York. He checked into a hotel every night, expecting the arrival of the General, Mary's sisters and brother, Willie. James supposed they were remaining quietly at the fashionable resort in Newport, Rhode Island and enjoying themselves, but he planned to go down the next morning to inquire of Mayor's Visitors for their whereabouts. In his letter, he noted, "The General's financial status must have improved."[25]

That night, James attended the opera and heard "The Barber of Seville." The next day, the Jesups arrived and James reported, they had met the Fisks while at Newport and had become quite intimate acquaintances. The General would return to Washington, but the girls would remain in New York to the last of September, with whom, he did not know. In the meantime, he took his meals with her family every evening and spent the evenings with them. The General had meetings with the Governor and other important figures. At the same time, Colonel George Croghan and General Zachary Taylor were fighting in the Mexican War.

Six months later, in New York, March 1848, James, in his efforts to secure a ship assignment for a voyage with cargo to Europe, was met with no success. His efforts eventually did pay off and he received ship appointments. His letters of despair changed to hope of better things to come and slowly, he was able to pay off his debts. His strong desire to provide a house and comfortable life for Mary would not go away, but he didn't give up, saying, "In the midst of this crowded city, I am in a desert until I know you, my Mary, almost any place was a home for me but now, I have but one, which is in your true heart."[26]

From the American Hotel, November 19, 1848 he wrote that since he had been employed in the steamship business he was convinced that it was almost impossible to make headway in it without some *capital* and he had none, but had some promise of it. "I sometimes get so vexed with my disappointments that I feel like quitting it entirely and then again the idea of going far away from you on some distant station renews my exertion, but again, my dearest pet, Oh! how I hate that word 'fail' in anything I undertake."[27]

James started for home on December 4th. He had probably reached home when news of the death of Mary's two uncles, Dr. John and Colonel George Croghan arrived. William Croghan was still in England when the news reached him.

While at home for two months, James Preston Blair loaned his son James $10,000 to start a shipping business in partnership with the Aspinwall brothers in California. He left New York on the steamer *Falson*, February 2, 1849. At the same time timbers for three ships were sent by sea to California around the Cape of Good Hope and were reassembled in California, mostly by his own hands. He left New York on the steamer *Falcon*, February 2, 1849 and sent Mary documents of evidence of their property in three steamers and other material of the *Enterprize* before departing for California. It was very important to acknowledge the receipt of the same in her next letter in case of any accident to his self.

Three hundred passengers were aboard, including a large party of army and navy officers. The confusion of putting baggage aboard and getting the three hundred settled was very wearing and tiring, but he would have a long time at sea to rest and plan.

CHAPTER THREE

California

Mary's concern about sickness and medical attention for her husband was put to rest when James informed her that a surgeon of each branch of the Public Service would be on board together. Before the steamer *Falcon* sailed out of New York, James had been so entirely overrun with the business of his and Mary's *Enterprize* and in the last few hours he was so exhausted from inspection he had neglected to write. It was a relief to get to sea for some sleep since he had been up every night until 3 o'clock. The confusion of tracing and checking three hundred passengers and their baggage made it almost impossible to get paper work done.

James sent Mary evidence of their property in three steamers and other material of the *Enterprize* before departing for California. Documents would be revised and renewed to James for signature in California, but until then, returned one of the originals of each to Mary.[1]

Before reaching the Isthmus, some old salts on board were warning him of the "Nasty Deep" or "old Monarch Neptune" and his temper and to beware when something begins. It had been five months since James paid him a last visit and it was doubtful how old Neptune would receive him since James had always been respectful, he may consider him a favorite.[2] He mentioned his recovery from the indisposition which troubled him just as he was leaving home.

Before James left his father's beautiful country estate "Silver Spring" just outside Washington in Maryland, he asked his father to build a small cottage on the Silver Springs estate for Mary and promised to repay for its cost. Francis Blair was very happy to do so, and wanted his daughter-in-law and little Violet close by. He started the building of the cottage, immediately.

The steamship James sailed on arrived in Panama on February 14 and rejoicing took over when he received Mary's affectionate letter of the 30th with news that his little flower and Mary, by God's goodness, both would be in James's sight in less than a year. James and some others had formed a party to cross the mountains together. It took seven days across the Isthmus which allowed him opportunity to make a passing examination of the country. It was the richest growth of spontaneous vegetation than he had ever seen in any of his travels about the world. The hot sun and great fall of rains seemed to exceed any part of the earth. Man could not enter any part of it (except the mule), without the axe and the knife.

A letter to Mary would be dropped in the U.S. mails at Havana on the English steamer due from the South. "Tell the General that everyone with whom I have conversed on the subject of my enterprise tells me that if my material arrives, I'll most undoubtedly make a great deal of money and after, my darling, I'll be able to maintain you in that position of comfort and ease which you have always been accustomed to."[3]

Steamer *Falcon* reached San Francisco, April 1st. It took eight days from San Blas, Mexico and up the coast which was very calm and pleasant. In fact, James had never passed over so much salt water with so little wind and calm sea (like a pond) in all his travels.

Letters from James Blair describing the San Francisco scene upon arrival of the "gold rush" or "want to get rich Easterners" is a vivid historical account of the settling of San Francisco and the beautiful love letters sent to Mary are very rare, indeed.

James wrote to Mary that it was almost impossible to realize the cost of labor and land in this town. A house in town about the size of Mr. Shomacker's in Silver Spring rents for $7,500 a year. Common labor in the street gets 10 to 20 dollars a day for loading and unloading carts. Provisions were, however, very cheap with average advance of about one third of the New York prices. There was however, a great want of servants; it was impossible to get them at two hundred dollars a month. A good cook could get three hundred a month. The cost of living was very high and the least he could get along with was $5 dollars a day. Captain Folsom had been very kind to him and in Folsom's quarters he continued to sleep.[4]

James had begun to fear that Mary would think him stupid as well as getting to be an old man. He had a better outlook for their future.

Captain Folsom informed James that he could take his invoice of his arrangements for transportation into the market and sell them for $100,000 dollars at least. James believed it was possible to make about "50 or 60 thousand" that year if he succeeds.[5] He would sell all his interest and return home. That would be the only time in his life that he would have an opportunity to make any money, therefore his whole effort would be turned to it for one year from the time he left home. The Bay and the neighboring rivers were beautiful waters for steam navigation.

Mary received a letter from James informing her that she had no idea what privation the ladies who had accompanied their husbands had to suffer. No servants in the houses and but few who would wash their linen at $8 dollars a dozen. Therefore, he rejoiced that she was at home enjoying our little Violet. Mrs. General Smith and several other ladies had arrived, but a month since had returned home by the next steamer. In fact, every lady who might come out there would return again with great hustle for the country was really nothing but "bubble, bubble, toil and trouble" as far as the ladies are concerned. The officers attached to the squadron consumed nearly all of their pay for fresh provisions and vegetables. The whole community was perfectly absorbed with the idea of making money.[6]

Those with whom James discussed his business venture were convinced his fortune was much, if his steamers and lighters arrive out safe. Yes, if even one of them arrived. The next week, James made a trip up the rivers to examine the parts for touching with his boats and returned again after ten days for giving Mary an account of his travels up among the gold miners. One of the miners had come "in" so that Captain Folsom could take care of 100 lbs of gold for him which was the result of his last trip to the mines. He had been absent about three months.[7] James informed "my darling" that if he should have good luck in the safe arrival of his boats, she would have it in her power to be charitable to the needy and enjoy every comfort she would wish in this world. Mary's "cottage," garden, and library she should have if God favored his exertions. To see Mary happily enjoying these, would make him perfectly happy.

In April, he had been up the [unintelligible] Bay and was employed on a survey for 15 days, accompanied by Captain Hammond and Sherman of the Army. Each one had received $500 in cash and six choice lots in the town which had been laid out at the termination of the survey which were selling in town at $150 each. If they get up as high as $200 the next week, he would sell half of his and remit the money home to Mary for he felt that she must be in want of it. "But, I hope not much for nothing grieves so much to know that you wanted anything and that I had not supplied it in time for it. This is my great aim and pleasure to give you every comfort and pleasure that this world can afford."[8]

The $500 which he made in the survey was invested in a schooner of 150 tons and it was employed in transporting freight and passengers up to Sutter's Fort and the town of Stockton on the Sacramento River. He was expecting to do good business and planned to go up to see and learn the character of the rivers and the business to be done on them.

A melancholy death occurred—a Mrs. Simmons who was one of the ladies who accompanied them out in the steamer *Oregon*; the "Poor fellow, he looks broken hearted. I never saw a man more devoted and in love with his wife than he was. He was all attention to her wants throughout the voyage."[9] James' heart was touched with much sadness because that was rough country for men and a real purgatory for ladies.

He was glad he didn't bring Thomas along with him since he couldn't afford the wages. Folsom's sofa or camp was his bed, using a bucket of cold water each morning and clean underclothes to complete his toilet. James wanted Mary to write him and teach his little Violet what to call her father and tell Ellen to take good care of both of them. He would remember her even at the end of his days.

"Poor Charlie Wilkins has lost all honor by committing several thefts to a considerable amount of money from several officers of the Army there." He had drank and sunk to a low level of degradation and had gone to the mines. "Poor fellow, I pity and feel for him to the bottom of my soul. He had many fine qualities, both of head and heart, but no balance of mind and no power to rise within himself and will anything."[10]

James cringed to think the miserable life it would have been for one sweet sister, Lutie, because he felt if she ever loved a man, she could follow him through the darkest shades of life, but dishonor would have broken the heart of her high toned soul.

In June, James received three months of letters from Mary and on the returning steamer sent her the amount of his earnings in California. Some gentlemen had paid him for land and hydrographic survey which lasted fourteen days and since that time, he had made $1500 for making reconnaissance into the interior and taking large vessels up the Sacramento. The last, he had invested by purchasing one third of a large schooner, hired employees, taking passengers and freight up the Sacramento and Suisun River. He continued to write to Mary of his desire to make her, "mistress of millions."[11]

He sent her three deeds for lots in town which were given to him by Colonel J.D. Stevenson for a journey of Suisun Bay. He took them out in his name, Mary's, and Violet's to try their luck in the great lottery *in the development of this wonderful country* and may be worth $10,000 in two or three years. They only cost him two days work. He took up another large vessel and joined General Smith at Benicia on the Bay landing. He was an invited guest of the Party and they made a grand tour of all the mines.

A great many persons with familiar faces and names were arriving and the 'strange feeling' was giving away to "in U.S.," again. He had not heard anything from Charlie Wilkins, but thinks he is up in the mines somewhere, digging, drinking and gambling, sunk so low that there was scarcely anything else left for him to do. James had one of those short spells of sickness, which Mary had witnessed before. He was only down in his back for two days.

Boats leaving New York would be on the ocean for six months before arriving in California. James was waiting for his boats and upon arrival would establish an office in a building where he would be settled with fixed habits. He sent three deeds for lots in town which were made out in Mary and Violet's names and thinks they will be worth $10,000 in two or three years. He was making every moment count and only two days of labor at a time when not much to do. By the next steamer, he sent gold specimens and Ellen (Violet's nurse) also received two natural gold drops for ear rings.

In July, again, James sent Mary two drafts for $500 each and he bought others with gold dust. One draft was sent "for what you may want. I have directed to be used in constructing that little cottage near my Father's for you and our Violet that you may always seek the country when you are with your friends." He wanted the General, sisters and brothers to spend the summers there. After all, the General had sheltered his wife and daughter for the greater part of the year. It gave him pleasure to know that Mary and Violet had established themselves at his parent's home at Silver Spring. Nothing gave him more pleasure than for Mary to see what he had always told her, that his father and mother love her so much.[12]

On the next steamer, James sent Mary, by a Mrs. King, specimens of native gold. He intended them as ornaments for herself, Violet and Helen as Mary thinks most appropriate. To Ellen, he sent more native gold. He wanted a small broach made for his Mother's shawl and strong enough for her thick shawl. He confined himself to the bold and hardy service as a pilot and even when his boats arrived his attention was drawn entirely in "developing this country by the power of steam and active business which could be so entirely professional."[13] Maybe some day, being a pilot will be looked upon as more professional than the work of his two brothers, who would return to the bench. Perhaps he will be viewed differently at home.

James enclosed (a newspaper clipping) some interesting news that no doubt would surprise those back East and he hoped would gratify because it was some hope of reform. Charlie's wife, poor girl, knows nothing of his villanies in the lower Country and I hope she may never. Things are known to a few afficers of the Army only. He had been elected to a high office of the District under the Mexican or Civil Law and has now a complete reform.

Married

On Sunday, the 28th of October, [1849] at Sonoma by the Rev. Mr. Woodbridge, the Hon. Charles P. Wilkins to Miss Emily C. Peterson, both of Sonoma.

The influx of the "forty niners" was becoming a great bother. To the new-comers or arrivals, James had the reputation of having been everywhere in the country and knowing everything about it and "you know that I am a good natured old man with a bald head. Never was a poor uninteresting witness more searchingly cross-examined than I have been of late your letters, my darling Mary are the only glimpses of happiness that I have in this busy dusty & disagreeable place."[14]

He continued to send more money home and wanted to call their cottage "The Moorings" which would be very appropriate as far as he was concerned.

The Moorings is a sea term for home, where it is snug and comfortable. When a ship is at her moorings, she is said to be quiet, peaceful and happy. When a ship seeks her Moorings all labor, troubles, dangers of the seasons ceases and all on board are contented, quiet and happy. It is used by the Poets and Romancers who have written of the persons and things which pertain to the sea. Therefore, my dear Mary, let it be called 'The Moorings'.[15]

December 31, 1849, James received five letters of August, September, October and November from Mary. He could not return in January as he had planned since he was unable to find a purchaser for his interest, especially one whom he could credit so large amount of money. The property was very valuable, much more so that James' highest estimate even reached, for instance, the receipts of the little steamer *Captain Sutter* had been $44,000 in six weeks running and she is the smallest boat he and Mary have. The *El Dorado* went into operation ten days later and her receipts doubled that amount. The expenses of running either of them never exceeded $5,000 a month.

He was so sad at the thought of not seeing them as he had planned, especially after receiving the daguerreotype which made him so sad and yet happy, every time he looked at it. Happy, because he could see both in his imagination. Mary holding that little hand still to keep it from spoiling the picture, and then, again he was miserable, sad and unhappy to feel that he was spending the prime, the joyous part of his life away from her to whom alone his heart owes allegiance. He longed to be near them and had no ambition to be great.

General Jesup had been spinning his usual yarns, obviously. James wrote to Mary that the Laurel Crown rests in purity and glory on the head of her Father and hopes and trusts that General Taylor will have honor and honesty enough to give the highest honors of the Army to Mary's father which he had so long deserved from the hands of others, and due from the Republic if he has the opportunity. "The General has done a great deal of the *thinking* for President Taylor in his successes of the Mexican [War], if he could only look to sources of it, he would plainly see the General."[16]

Winter in California was a horrid place to live, especially when James could see nothing about while attending his outdoor business but drunkenness, distress, mud and rain. The whole face of the earth was mud a foot deep as a result of six to eight hours every day of the week of raining. If he lives, he will be with Mary in June but, sent a letter by a "special agent" which he wanted read to his Father. From the letter, she and his father would be able to gather an (unintelligible) of his business transactions and at that time, the amount even astonished him, when he comes to think of it. He quiets Mary's fears in regard to his transactions in business for he does nothing but a Jacksonian Business (specie payments) to the last dime. The experience and its results of giving, receiving, enduring, indorsing or accepting notes of individuals or corporations of both his father and father-in-law are constantly before him. He had never endorsed a note and had rather loan the money. Remittances would always equal the balance on hand because he buys or sells for cost and therefore pays cash. He sent more lumps of gold home with specifications as to what ornaments were to be made and for whom. The large pure lump of virgin gold with a coronet or rosette was for his dear Mary.

Back in Washington, Julia Jesup was ten years of age and presenting some degree of misbehavior for those around her. James wrote that he was glad and rejoiced to learn that Julia was under Mary's management and she had become manageable, good and considerate of others. Even in later years, it seems that Mary had the role of "Parent" for the Jesup siblings. James welcomed the news that Lutie and Janie had resumed their sway over the gay circles of Washington again, after Janie's health had improved. James had received a year's leave from the Navy, but recently had perceived that the Secretary of the Navy disapproved of officers in the Navy seeking mail steamship service. He concluded that any hope in search of command of some sort was utterly hopeless in the Navy. Twelve thousand dollars was sent home of which seven thousand was to be paid for investment in the New York line of steamships. Other investments were made in railroad stock and a loan to the *Globe* newspaper at 7 percent, per annum.[17]

Being involved and learning from James about managing business affairs served Mary very well, in later years. James was her strength and pillar in preparation for weighty problems and decisions which would come her way. Mary had learned that one had to stay on top of business operations and stand her ground. Her stamina and fortitude came to the forefront when she, alone, had to make her voice heard, and acted upon matters at a time in history which was referred to as a "man's world".

If the Navy Commodore was not to permit James a return home by the first of October, he was planning to resign and come home. He had continued sending large amounts of money home and had bought a lot on "the square" [Lafayette Square in Washington D.C.] where plans were underway to build a town-house. He figured a house in town and a cottage in the country with $6,000.00 a year or more to live on should make a long life together with books, flowers, and finer things of life. An agreement had already been made for sale of one-half of his shipping business.[18] He detested that horrid strife of making money in which many who were thought to be honest and true, had lost their house and character. The plans to depart for home on certain dates were always changed which seems to have strengthened the desire even more. He always ended his writings by expressing a driving desire to come to Mary and his baby with "if God grants me health and life to get there."[19] It had been a life of hardship and vexation with nothing but prospect of long continuance even to the end of his natural life. James also said he had endured more mental distress in the last seven months than in all life previous.

The celebrated architect, Mr. Ranlett, had been working with James in drawing up the plans and specifications, with New York estimates for the house to be built upon the President's Square. The whole property would increase in value as the Republic grew at the Capital. He had no doubt that Washington would remain the Capital of the Republic for many years to come and felt certain that distance was then completely annihilated by steam. It makes but little difference where the seat of government will be.

October 15th, James was surprised to receive from Mary two letters which had been written in August informing him that she was coming to California to be with him. A short notice was sent to "My dear Mary, If you have not left for San Francisco, do you think that you deserve a letter for me. I will there fore, scold most furiously in my next letter, if I do not find you on board the Tennessee this day, one week."

The Blair family had been behind the discussion to send Mary to James, in fear of him leaving a thriving lucrative business and also, resigning from the Navy. James would have Mary's heart and home with him.

San Francisco
Nov. 15th, 1850

My Darling Mary, I pushed on board the steamer "Tennessee" in a great state of excitement, expecting to find you on board, but I was doomed to disappointment and my pet, you have no idea how very bleak I felt. If you do not arrive in the steamer California, I shall quarrel with the Fates terribly.

I have a pretty little cottage built house for you to live in and in a very disrable part of the City. it is all however, subject to your approval. I am anxious for your arrival on the 21st. James Blair.[20]

Mary arrived safely and sent a letter to her father, General Jesup, informing they had reached Chagres, a river rising in central Panama and flowing to the Carribean, the night before. The weather was fine and in a few minutes they were to go up the river. Colonel Fremont and Mr. Rodgers [Augustus?] were very attentive and did everything to make Mary Blair comfortable. She was ready to leave, but wished the family back home could see the town. The country around and the old fort was beautiful; however, she found the inhabitants and houses were miserable.

After Mary's departure from Washington, Serena and daughter "Tinie" arrived at the Jesup home to spend the winter with the General and three daughters. George Croghan had died two years earlier.[21]

Very soon after Mary's arrival in California, her family received letters giving descriptive accounts of her new surroundings and how happy she and James were in their own home. There were few people in San Francisco who were making fortunes, but it was so terrible to think of the immense number of people who were starving. It was not like Washington where the hungry could be fed from every door. She thought her husband's success was due to his kindness and goodness.

During the day, James was at his office and Mary would take Vivy out walking while Mammy washed and attended to her work. In the evening, Mr. B.[Blair] would read while she worked, except when they had company. James celebrated her birthday with a small dinner and one of the guests was Mr. Fox, who was engaged to Miss Woodbury. He had been to see Mary several times. The Fremonts lived opposite to them and their Lillie came to play and walk with Violet nearly every day.

Sometimes she wrote as if in her diary, "… A terrible fire broke out. I have just given Mr. Blair his pistol that he may go to it. 2 o'clock—night—Mr Blair has just returned, the fire was put out sooner than anticipated—he lost nothing. By the last fire, he lost $10,000. Folsom lost tonight, near $20,000."[22]

15th February 1851, James gave Mary a beautiful new piano for New Year's Day. She was anxiously looking for letters from home and hearing from her family. Mr. Flandin (who attends to Mr. Fremont's affairs) returned to Washington on the last steamer, promised to call and tell how they were fixed in California. Little Violet went very often to see "Aunt Jessie" (Fremont). Colonel Fremont was still in San Jose and no one knew how the election was to turn out, but "Mr. Blair" went down to attend it. The weather was delightful, but winds were so high and dust so dreadful, ladies and children could not go out after 10 A.M. each morning and it was very cold. Little Violet would pray for those back home, half a dozen times each night and insisted on telling Grandpa Jesup that she was coming back to eat some of his "corny-bread" and he will say she is a good girl for coming to his house. Mary was anxious to hear from the boys at West Point.

In May, there was another dreadful fire in San Francisco and the Blairs had a house full of friends who were burned out. James lost very little; even though his office was burnt, he managed to save all his books, papers and money. A letter from Mary in which she still seemed to be in shock reveals how powerful a fire can be. The many houses, even the streets burned and so many lives were lost. Frances P. Blair received a paper giving an account of the fire. James and Mary were very near burned out due to the rapidity of the fire sweeping over the town. Their escape was a miracle.

The Fremonts' cottage burned, but the next day she was comfortably fixed in another. Mrs. Fremont had a very fine boy and Mary sent Mammy over to dress and attend her every day as she had no nurse. The news was that Mr. Fremont stood very little chance of being elected that fall.

The Blairs' house always had bouquets of the most beautiful wild flowers. The mornings were quite warm, but about noon the winds would commence to blow such cold air, that large fires were necessary to keep warm. Letters from home made the Blairs long to be back when they read how beautiful Silver Springs was with everything in bloom. "There is nothing to see in San Francisco but, brown looking hills without a single tree. The climate is like Indian summer back in Washington. There is a beautiful view of the harbour from our house and we watch the shipping vessels coming in and going out"[23]

In October, James' business partner Aspinwall's conduct had annoyed him very much, but fortunately he had secured himself and would lose nothing by Aspinwall's failure. Mary was anxious to know what the people in the East think of the Cuban Affair. Violet had been taken to the dentist to have plugs put in two of her teeth. She was very proud of her "California Gold" and would have a new coat for the winter, made of the white marina sent by her Grandmother Blair which had been embroidered all over with white silk thread.

March 1st, Mary wrote about her new baby boy, named Jesup and weighed ten pounds and four ounces.[24] The bureau was covered with flowers sent by friends, but she regretted that it would probably be a year before they could return home, then they planned to go to Europe.

Prior to the spring of 1852 had been a troubled time for Mr. Blair. He had a difficulty with some men which resulted in a duel with one of them (a

Mr. Peachy). Friends on both sides tried to settle it but did not succeed, and Mr. Blair sent a challenge to Mr. Peachy. The fight didn't come off for three days after and Mary had endured a dreadful state of anxiety and suspense. She thanked God that no one was injured and trusts they may never know such sorrow, again. Everything had been satisfactorily explained. Dr. Bowie was Mr. Blair's friend and Dr. Hastings his surgeon. Every one seemed to feel so much more for Mr. Blair as his wife and children were there and Peachy was an unmarried man. After Mr. B. was safe at home again, his friends sent great quantities of flowers to Mary—many from persons whom she had never seen, but were his friends. Their little Jesup had been very sick, but had then recovered. During the time of stress, a wet nurse for him was necessary for a short time, but afterwards, the Dr. allowed Mary to nurse Jesup.

Mary heard from home that Buchanan would certainly be the next president. She hoped the news was wrong. She also heard accounts of Julia's beauty which pleased her very much. There was a fear that those at home would forget Mary and family if they remained away much longer, but the Blairs made the best of the situation by making drives in the country. One special place was out to Bledsoe's Rancho; they would drive for miles over the most beautiful wild flowers and there must have been fifty varieties and very fragrant. Some of the hills were covered with nearly ripe strawberries.

In June 1852, San Francisco was growing so fast, an immense number of fireproof brick buildings had been built so they can never have such fires as they had the year before. Chinese were the best and most industrious foreign population in California. The next dry season, she was planning to leave on the Isthmus, expecting business to be settled by then. She was so anxious to see those back home, but couldn't leave James alone. If he goes before his affairs are settled, no agents could be trusted for settling them.

Mary had several attacks of illness since the birth of her little boy. The Blairs had gone to the Rancho to remain until the summer winds ceased blowing. Her health had improved and Mr. Blair's rheumatism had almost left him. A return to Washington was planned for the winter. Every day, Mr. Blair drove Mary, children and Mammy, sometimes into the mountains where the air was so fine and beautiful. James amused himself teaching Mary to shoot. Someday, Violet planned to stop in the mountains and dig some gold to take home to everyone of his people. The wife of one of Mr. Blair's best friends, Mrs. Randolph, who knew the Lees very well, was staying at the ranch and she was a very dear friend of Mary. Gossip was not left back East, it was alive and well in San Francisco. Another friend, Mrs. Jones, had left a few weeks before. Mary hoped the reports which were in circulation (San Francisco), about Susan Jones were incorrect, but everyone seemed to have heard it.

The Blairs planned to return East in October, but James was afraid to take the family while the cholera was so bad at the Isthmus. They were planning to build their future home when leaving California and to Mary it would be without a single regret. She could be happy anywhere with her husband and children.

Mary and children did return home that fall, but without James. The next letter was sent by James from California to Mary, dated December 1, 1852. He had kept the diary of current events of every day as he had promised and had been very busy selling the furniture out of their house and building an extension on another little house where he intends to live. He had been very busy with business and had not a permanent place of sleeping, "first at one place and then another, sometimes on board the Action with Alden and then at other places, not often in my little house, which is crowded with furniture and smells strong of fresh paint."

James was making use of Peyton's desk and having a difficult time writing the December 1st letter, since old John Parrott and Peyton were both talking to him at the same time about the MaConfry and Peytons amicable intentions for the young ladies. Peyton was urging old John to purchase a lot to build on one of Mary's lots on Stockton Street so that he may be conveniently next door and so drop in neighborly and feel his way to a secure position in the upper ten society of San Francisco. They had been pleading hard to get a piece of Mary's lot but Mr. Blair had refused every proposition they had made.

James' man Thomas and nurse Mammy had returned with Mary and children. He was completely alone, so sad and wrote to Mary, "My darling Mary, you have no idea how much I have missed you and my babes since you left me. I dread to go home at night. Everything is so sad and silent there. I moan aloud and sometimes whistle and sing that I hear my own voice. As soon as I get settled in my own little house, I will write every evening."[25]

December 15th: James had not written a letter in two weeks. Only the night before, was the first evening he was able to get his "little house" free from confusion which had prevailed there, due to rain and Marie being sick. Everything was crowded into two rooms. Order prevailed, at last, but still he could not help feeling the truth and force of the stanza he had just been reading:

Home Is Where There Is One To Love

Home's not merely four square walls, Though with pictures hung and
  gilded;
Home is where affection calls—Filled with shrines the Hearth had
  builded!
Home! go watch the faithful dove, Sailing 'neath the heaven above us.
Home is where there's one to love! Home is where there's one to
  love us.
Home's not merely roof and room, It needs something to endear it;
Home is where the heart can bloom, Where there's some kind lip to
  cheer it!
What is home with none to meet, None to welcome, none to greet us?
Home is sweet, and only sweet, Where there's one we love to meet us!

[Charles Swain]

Everything was then quite comfortable about him, in fact, the house was snug and clean and pretty, but yet, it did not make him happy, nor did he even smile to those who were kind to him. He reflected that he had no right to complain, his career had been so changed in life, for the better. What officer of the Navy had the prospects he had for a happy life at home. It was his duty to stay there until his affairs were settled and fixed right. He would never forgive himself if ever Mary and his babes should want and was not able to give and in abundance too. He would be sad, but would not moan.

His work was his only relief and to that he devoted himself. An agreement with three others had been signed to continue Combination on the Stockton boats for six months longer. Three were then in control of the business and they had purchased the steamer *American Eagle* for $35,000. Their profits had already paid for two-thirds of the cost.

James had recently heard Kate Hays sing several times. He thought she was a charming ballad songstress and would be a great favorite with his father as she sings Burns' Scotch songs with the poise and sweetness of a native lassie—a fine actress and yet has the unaffected bearing and manners of a "Lady".

James had received a letter dated November 27th, from Panama and learns that Mary got safely as far as Cruses, but he would be in fever until he hears of her arrival home. He had a severe attack of rheumatism in the shoulders and back. Stockton Street was in a quagmire from landfills to the seashore after twenty days of raining. All the bridges were washed out, cutting off travel and communication. The cargoes arriving from the east were choking up the warehouses with supplies and miners in many parts were starving. California "is perfectly adrift."[26]

James and two other business partners had just signed another agreement, forming a combination for all boats on the Stockton Route for six months.

James hastily sent a letter January 14, 1853, informing Mary that he expected to be with her before she received his letter. It was a short stay, but he managed to get Mary and children settled in the Moorings and attended to business before returning to California. He was back in San Francisco by July 31st.

Mary's house in San Francisco on Stockton Street was then worth thirty thousand dollars and renting for $175 a month. Janie had been married for a year and her husband, Augustus Nicholson, and his father were also in San Francisco.

After James returned to California, he had 15 days of frightful suffering-relief from cupping. Bowie had cut and cross cut his entire back.[27] He was soon up and walking and wrote that he had walked home from church with Tinie Croghan last Sunday. "She is quite pretty and very agreeable."

September 15th: He was trying to get his affairs in order but, impossible before March or April, 1854. James Blair was filing a cross bill of complaint in D.C. against Aspinwall. At that time, he was concerned about "never been away in your [Mary's] time of trials."

September 30th: Mary, Lutie and children had been at Columbia House, Cape May since early August. They were back in Washington D.C. in December, waiting for the birth of Mary's baby. A letter arrived for Montgomery Blair from California, with the breaking news of his brother James's sudden death.

Private

Dec 16, 1853
San Francisco
10'clock P.M.

To General Jesup, Washington

Dear Sir,

... of the arrival at noon of the mail from New York, the departure of the Steamer of today is delayed until 3 p.m. This affords me opportunity to communicate the result of the autopsy, which Mr. Blair's friends were united in thinking proper to order. I write supposing you will simultaneously, with this review a letter which I have already mailed for you upon the dispensation which must be announced to you by the Steamer. At present, I will merely state the result of the post-mortem examination. It was conducted by Dr (?) the most eminent Pathologist of San Francisco. In the presence of Drs. Tripler A.S.A—Grey—White, Gibbons—Mr. Casseby (?) & myself. They discovered that there had existed probably for some years an aneurism in the descending aorta about an inch below the arch wherein there had grown an abnormal adhesion both to the spine & to the back & upper part of the superior lobe of the lung on the left side. This aneurism had burst (probably on Wednesday Night, when Mr. Blair first taken sick) the blood had permeated the lung & forced at length a passage through the lung into the cavity of the chest. Mr. Blair died therefore from bleeding caused by the bursting of the aneurism. And sooner or later this abnormal condition of the aorta must have produced death despite every means and effort which medical men could have devised even had they known (what was almost impossible to discover) the exact nature of the disease.

So sudden was this lamented disease—so hale & well in every appearance was Mr. Blair up to the first moment & so few men present when he breathed out his life that it was very necessary for the satisfaction of his friends here that this autopsy should be had. Construing your probable opinion in the matter by my own dealing, I could not but decidedly assent to the general wish, specially when I considered the distance by which Mr. Blair was separated from his family and the time that must elapse before you reading the sad intelligence. Yet, it has been the effort to have the matter so privately conducted and so communicated to you that Mrs. Blair may remain in ignorance of the fact, or be informed according to your judgment in the premises.

For your sake, dear sir, & for my own amazement & deep sorrow in sympathy with you & yours. But be assure as from time you shall hear, (as I know you will) the reports of the esteem, confidence, respect & love which Mr. Blair was held by his many and warm friends & associates. You will be more & more soothed by the thought that he was not cut down until his full manhood of vigor, honor, virtue & affection had been nobly developed before the world as witnesses.

I am Sir, with great respect, truly yours

Chr. B. Wyatt

P.S. Allow me in a word to renew my request that you will command me in everything whereby I may possibly send Mrs. Blair & your family.[28]

Mary gave birth to James Blair's last child on 26th of December 1853 — eleven days after the child's father had died. The daughter he would never see was named Lutie *James* Blair and would always be called, "Jim." Lutie James Blair was sixteen days old when the news of her father's death reached Washington D.C.

Francis Preston Blair sent "Dear Mary" a note expressing his inability to see her the day of receiving the news, but he was too depressed and dislikes making a spectacle of himself in the city. Upon Montgomery receiving the sad news, the next day Preston Blair wanted to find Mary and communicate with conversation over the necessity of making an instant reply to be sent by the return steamer regarding James' business and fortune made in California, "...it might be well to forward a certificate of Dear Mary which prevents the rule of her affairs from being communicated (?) to her."[29]

17 Jan 1854 Silver Springs

My Dear Mary,

I have suffered in the loss of your husband, the greatest calamity that has ever befallen me. It is a loss that can never be made up to his Mother or to me. You and your children are to us as taking his place afford the best consolation to which we look for relief.

In this we trust that you will live with us or near us. If you choose to live with us, you shall have the best room in our house as your own and the largest as your nursery. This will enable you to make what you please of the Moorings. If you prefer making that place your home . . . of the years in carrying out the . . . of your husband . . . city conveniences and comfort, will be open to you to make up the deficiency of your place a country residence. Whatever you settle upon as the best suitable arrangement for yourself, I will contribute to accomplish.

I have telegraphed Frank telling him that James, as his last words with you had pointed to him as the one on whom you should rely to take his place in the management of your business . . . more . . . man couldn't have been selected for he was devoted to his brother as more energetic to give affect to his wishes. I expect he will soon be here.

My wife and myself, I am confident you will find ... kindred and our own children, relatives are willing to comfort and assist you as ... reason in blood.

I will not intrude my grief upon you, until you are able to bear your own, which I hope will be in a few days.

Your Aff father

F.P. Blair [30]

Silver Springs

Dear Mary,

Your kind note is very satisfactory and was consoling to me. All is right now. Mr. Dandridge called and supped with us last night and told me that James had authorized him to do what you wanted and he . . . at once.

Pray give orders to Mr. Goodin ? to take the proper steps to . . . an injunction.

Yours affectionately

F.P. Blair [31]

How many desolate hearts and homes there must have been in Washington D.C., California and many other places upon receiving the news of James Blair, who would never again be with them.

The inclusion of the preceding letters was considered important for the reader to know who James was and realize what he meant to Mary's life and lost much too soon. He reached his dream of making Mary a "millionaire madam" but, how empty the following sixty one years must have been without James. She never remarried. She had learned much from James, his inherent knowledge and perception of what really mattered in life. He possessed wise perception of others and appreciated the goodness in them but, he exercised fairness, firmness and cautiousness when needed in business dealings. He liked and enjoyed the company of others and was a gentleman to the end. If he had lived, he also had plans to "develope this country by the power of steam and active business which could be so entirely professional."[32]

One month after receiving the news of James's death, the brother, Montgomery Blair, sailed for California to settle his estate. He was surprised to learn how James was loved and revered by so many. James had been so generous and good to others. There were countless uncollectible notes ranging from small change to thousands of dollars. The lawsuits brought against the Aspinwalls and John C. Fremont were settled in favor of James Blair's widow. Many weeks passed in California before Montgomery finally commenced his sad long journey home with his brother's body and possessions put aboard a ship. It is possible that Serena, Tinie Croghan and the Wyatts accompanied Montgomery and James' body on the same ship which was returning home.

Five weeks after receiving the sad news from California, Lucy Ann Jesup married 28th of February, 1854, Captain Lorenzo Sitgreaves of the Topographical Engineers. He was a graduate of West Point and in the Creek War was under General Scott and General Jesup. In the Civil War, he was engaged a portion of the time on the defenses of Louisville. Lucy Ann Sitgreaves' letters had a Louisville hotel letterhead during that time.

CHAPTER FOUR

Widow of Lafayette Square

After Lucy Ann's wedding, James' mother accompanied Mary and children to an Eastern resort in the spring and were there for some time.

Mary soon had to make decisions on her own. It was one thing to have been left a wealthy widow. It was another to protect and manage her own financial affairs. Mary was fortunate to have some honorable men in her life whom she could trust. James had made wise investments while living and at Mary's request, Montgomery advised her on making more financial investments which also included property in St. Louis. As the years passed, nephew Blair Lee gave her sound advice while performing legal work for his Aunt Mary. The utmost respect and love were shown by Blair Lee in difficult times as if she was his own mother. She had engaged a very dependable law firm in California who made regular financial reports on her property in California. Mary's father, General Jesup, was someone else to contend with.

Dear Father,

I was surprised and hurt to find this evening that you disapprove what I had done in regard to the R.R. and I wish to right myself in the matter.

I told you when I saw you several days ago all that had passed. What I had asked of the company and what I intended to do if they did not assent to my wishes. I understood you to approve entirely of what I was doing as you said, to go ahead and get all out of them I could and you were quite anxious, I thought, that the bridge should be a good strong one and said, that what you had allowed them to do did not affect me for any damages I might require. I thought it my duty to do what I have done and granting to them the land which you guaranteed to them.

I cannot by bringing a claim for any damages to my property and appose in any manner your permission to go through the place.[1]

Mary Serena Jesup Blair (courtesy of Historical Society of Washington, D.C.)

Mary and her sisters missed the Jesup brothers who no longer were home in Washington City. In April, 1857, William Jesup was living in Western Kentucky and wrote to sister Janie Nicholson, thanking her for the seed she sent. He had been suffering from chills and fever for a few weeks past. He was telling Mary how beautiful the country looked at that time. The corn was growing, but he received the tobacco seed too late. William was wishing all would write as often as they could. It gave him so much pleasure to hear from home, "I seldom hear from dear Papa now, but I suppose that he has very little time to write private letters."[2]

The second Jesup son Charles Edward, admitted to West Point, brevet 2nd Lieutenant, July, 1858, was stationed in Fort [unintelligible] in California. He wrote to Janie that he had written to one of the family several times, and received very few letters from home. "This place will all burn up this summer. The thermometer had been as high as 103," and he was tired of soldiering. He paints a visual picture of the detestable settlement. "This place is a wash-basin bottom side up in the center of a large table. Imagine the basin to be a hill of sand and gravel without a tree or shrub of any kind, and the table to be a sandy place." Charlie was pleading for Janie to write more.[3]

In the 1860 Todd County, Kentucky census records, William Jesup, the oldest Jesup son was listed as: *Jesup, Wm. C., 27 years old, Born Washington City*. Most of his neighbors were listed as farmers whose value in property ranged from $600 to $8,000. William Croghan Jesup is listed as "*Gentleman*" with property value listed as $25,000.

Thomas Sidney Jesup died at Washington City on the 10th day of June 1860 (71 years old).[4] Charlie resigned from the U.S. Army in August, two months after the death of his father.

William Croghan Jesup died at Elkton, Kentucky 14th November 1860 (27 years old).

Charlie could have arrived in Kentucky just before William's death or immediately after William's death. Mary Blair received a letter from Charlie, Spring Grove, Kentucky in April 22, 1961.

My dearest Mary,

I have been spending four or five days with the family of Cousin Lucy's brother and found your letters here upon my return. I was very much pleased with my visit, for I was received with real Kentucky hospitality and then the family spoke so affectionately of Cousin Will for so they called our poor dear brother.

Oh, Mamie, you don't know how dearly our dear Willie was loved out here. Mamie, dear won't you please have your picture and the childrens' taken for me. I can't be with you, dear Mamie and to have your picture would be such a comfort to me. And Mamie get Janie to have her picture that [illegible] took of him and Jesup and Gus copied on Daguerreotype for me. Please have them sent to me as soon as you can. There is a daguerreotype here of Father but, I do not like it. I will not write any more. I will not go to Louisville, dear Mamie and indeed I have no desire ever to see any city except the one where you all are, my precious sisters. My dearest Mamie my love and kisses to the children and to Jami and Jesup, and for yourself my own dear dearest Mary accept the best love of

Your devoted brother,

C.E. Jesup.[5]

Marked on the outside of the above letter, "My last letter from my own darling precious Charlie." He died one month after the letter above was written on 22 April, 1861.

Once again, April 1861, Clark and Croghan relatives went off to fight a war, a historical event which proved to be the most horrible time in United States history. Fort Sumter was in the news and Robert Anderson was in charge.

May 8, 1861, a Blair relative Rebecca Gratz wrote, "May God watch over the Union and help those who labor for its preservation. Lizzie Lee & her boy left us this morning. She left her husband at Brooklyn on the point of sailing. She went home today, taking Miss Jesup [Julia] with her. Her sister, Mrs. James Blair has her family at Bethlehem for the season or until Washington is more quiet and safe. All able men of the family are drilling in the 'home guard'."[6]

John C. Breckenridge of Kentucky was a cousin of Francis Blair. Breckenridge became a general in the Confederate Army and toward the end of the War, Confederate General Early made a raid on Washington. At the same time, Breckenridge returning to Washington made the old Silver Spring Home (near The Moorings) his headquarters in the very library where he had been as guest with his old Kentucky boyhood school mate, Francis Preston Blair. He placed a guard over the house, protected it from looters and escaped the burning that destroyed Montgomery Blair's house "Falkland" on the adjacent hill. Miraculously, the Moorings escaped destruction.[7]

In Blair letters, mention is made of President Lincoln's buggy drives out to visit Silver Spring on Sunday afternoons. All these rides in the country ended when the Blairs and Mary witnessed the turbulent days following the death of Lincoln. Mary Lincoln was without friends in Washington and the family doctor prevailed on Mary Jesup Blair's sister-in-law, Elizabeth Lee, to stay with "that woman who was desolate and alone confined to her bed." Eliza Lee despised Mary Lincoln, but stayed with her until she was able to leave the White House. Lizzie Blair Lee was the only one to accompany Mary to the train to see her off when she left Washington D.C.[8]

Two years after the war, Washington was slowly returning to normal, and in 1867, Mary Blair and children were touring Europe in grand style. The young Jesup sister, Julia, had been living in Europe for some time and was waiting for their arrival at Havre. Mary was so pleased to see Julia, describing her as "very stylish and still beautiful."

Mary and children had prepared themselves for the trip abroad by studying and acquiring knowledge of European history. They planned to stay abroad for one year and enjoy everything that Europe had to offer to the fullest. Mary's diary is filled with notes about cathedrals filled with art, etc., which was of great interest, but most of all, she enjoyed the public gardens and fields of flowers in the countryside. They settled in and stayed in Napoli for their last months abroad.

Eighteen year old Violet Blair was a beauty who left a trail of disappointed suitors in Europe. They fell in love with her and some proposed marriage, but were rejected. A year in Europe was a constant round of parties, dancing and promenades for Violet. There were newspaper articles written about the American beauty. Meriwether Lewis Clark, Jr. (grandson of General William Clark), a lover of the thoroughbred, had

Julia Jesup (from the collection of Historic Locust Grove, Inc.)

Violet Blair (courtesy of the Historical Society of Washington, D.C.)

Jesup Blair (courtesy of the Henry E. Huntington Library and Art Gallery)

Luti James (Jimmie) Blair (courtesy of the Henry E. Huntington Library and Art Gallery)

Meriwether Lewis Clark, Jr. (from the collection of Historic Locust Grove, Inc.)

traveled to the British Isles and Europe that same year. He had a special interest in the sporting world of horse racing, in preparation for something big back in Kentucky which would be the Kentucky Derby at Churchill Downs. He followed the Blairs to Paris, Naples and Rome. The tall handsome Clark cousin was falling in love with Violet and became very possessive and jealous of other attentive suitors. Mary Serena began to listen to Lewis, Jr. who was making decisions on how Violet should conduct herself and took measures to keep all other suitors away. Violet rebelled by "setting Lewis up" for a proposal of marriage by teasing and pretending to be in love with him. She took great satisfaction in saying "no" to his marriage proposal and vowed, "Woe to the man who dares to interfere with me again."[9]

Violet had left a New Orleans gentleman waiting back in Washington, who was Albert Janin and very much in love with her. He became so jealous and concerned about losing Violet while she was in Europe, a Blair relative suggested he join her in Europe and he lost no time in following her suggestion. Violet and Albert Janin would always recall the happy memories of being together in Europe.

Six years later, May 14, 1874, Violet married Albert Janin, whose family of lawyers and mining engineers was well known in New Orleans.[10] Albert entered politics and operated a canal in Louisiana which was never a financial success. In the fall after a long honeymoon trip in New York and in the Great Lakes area, they were on their way to New Orleans. Violet and her new husband stopped for an overnight visit with Meriwether Lewis Clark, Jr., his new wife Mary Anderson and baby son in Western Kentucky.

Bowling Green [KY] Nov. 8th

My Own Darling Precious Mama,

We could not get off last night as the train only went as far as Nashville. You have no idea how I miss you, and how homesick I am already, in spite of the kindness of everyone. Of all the people I have met I prefer Dr. C. [Jonathan Clark's son]. He is such a highly cultivated and charming gentleman. It is such a pity that he should be buried in this place. I could see that he took a tremendous fancy to me. He brought his Dante and Horace and sat down by me for hours to read and talk them over.

Lew [Meriwether Lewis Clark, Jr.] and Mamie [Mary Anderson Clark] insist upon our staying with them when we come back from New Orleans, but I will be in too great a hurry to be with you again.

Mamie's boy is a pink and white angel with big brown eyes and golden curls. He is perfectly lovely. Old Cousin Lew [Meriwether Lewis Clark] was staying there and sends his love to you. Their house is very comfortable and quite large, but is not completely furnished yet. They keep the parlor as a sort of store room. George Clark [son of Meriwether Lewis Clark] has come out of the asylum well and is doing remarkably well now. An asylum is a good place for some people when they behave badly. Your own devoted—Violet.[11]

One year later, 1875, Lewis Clark, Jr. established the famous Kentucky Derby.

Mary's other daughter, Lucy James Blair, (Jim) married, in 1875, Major George Wheeler, engineer and West Point graduate. She accompanied him to his military posts in Denver and elsewhere in the U.S. and Europe. Their life together was filled with hardships, which ruined the health of both and the next few years, Mary spent much time in New York, caring

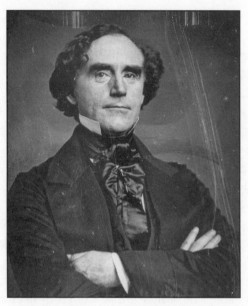

*Meriwether Lewis Clark, dauerrotype by Thomas M. Easterly,
1850 (courtesy of the Missouri Historical Society)*

*Abigail Prather Churchill Clark (courtesy of the Missouri
Historical Society)*

for and nursing the young couple back to health. Wheeler was bedfast for over two years and then had to use crutches. The youngest daughter Jimmie, who never knew her father, had an unhappy life.

Both sons-in-law had become indebted to Mary, and they failed in business. In the end, each one would call on Mary to endorse another note or extend another loan to save their greater losses. In the late 1890s, Mary had signed a mortgage loan for George Wheeler, valued at over $20,000. Mary had to seek legal help from her nephews, Blair Lee and Woodbury Blair when a mortgage company repossessed the house which had belonged to Mary in Washington D.C. on the "Square." At that time, the Wheelers were not speaking to Mary or Violet.

The Wheelers lived in a very expensive mansion in New York. Cousin Blair Lee, who was very close to his cousin Jim Blair Wheeler (Mary's daughter), visited the Wheelers at Jim's request. Jim wanted Blair Lee to help in getting the bedfast George committed to an army hospital as insane. Blair Lee reported to Mary about the visit, learning from their doctor that they couldn't pay their bills and the house they lived in was mortgaged to the fullest.[12]

Another stressful situation was going on in Mary's life at the same time. Mary Jesup Blair was the one that saved the Mammoth Cave for the remaining six original heirs who were all females. She invested a large sum of money in the Cave and in the end she received no encouragement from the other five female heirs. Lutie's daughter, Mary Sitgreaves, removed her mother from the family squabble by taking her (Lucy Ann Jesup Sitgreaves) to Boston to live. Julie moved to Italy and remained the rest of her life. There were strained feelings between Mary and sister Jane because Mary found it necessary to bring legal charges against Jane's husband Augustus Nicholson and others concerning Mammoth Cave.

On April 1, 1902, an awful bereavement came to No. 12 Lafayette Square. Fifty year old Jesup Blair, "the life and soul of the household", died of a cerebral hemorrhage.[13] Jesup was the one Mary depended on for assistance in Mammoth Cave business and made sure his mother was on safe legal grounds in Cave business by consulting lawyers. He loved the Moorings in the country and took great pleasure in making the grounds and flowers a showplace for his mother. Jesup inherited his father's admirable traits of kindness and everyone loved him. An example is when he came upon a black woman struggling with a basket of laundry, he stopped and went quite a distance out of his way to take her home in his buggy.

The following year after she had lost Jesup, Mary's daughter Lucy (Jim) died. The loss of two children plus, the struggle to regain control and saving Mammoth Cave for the rightful heirs was a task for only the strong. The most difficult situation was taking place with the son-in-law George Wheeler. Just before Jim's death, mortgage companies had repossessed Wheeler's property including the three story house on Lafayette Square, given to Jim as a wedding present. Wheeler was filled with anger and revenge when he learned Mary had been the purchaser of the house. He refused to return a broach of Mary's which James Blair had given her on their wedding day and in turn, had been given to Jim on her wedding day.[14] The most treasured item held by Wheeler was a family heirloom silver tankard that Wheeler refused to give up. The silver tankard had been in Mary's family for years and was given to the Wheelers for a wedding gift. The prized possession could have been given to Mary on her wedding day from her mother Ann Croghan Jesup. William Clark's grandson, William Hancock Clark, knew the story behind the silver tankard and put time and effort in an attempt to return Mary's silver tankard to her. Clark finally had to report to Mary, "The silver tankard was given by Wheeler to Nellie Voorhis to spite you, the day after he learned you purchased the house No. 930 Sixteenth Street, Washington. She resides with Mrs. Jeff K. Clark at 92nd St. NY and would return it to you if Aunt Sue is informed as to the same being 'your heirloom' and not Wheeler's."[15]

In the end, it was required of Mary to produce a document of proof confirming the origin of the heirloom tankard. It was highly unlikely that William Croghan, Jr. received any written document, upon receiving the Croghan family silver tea service which would have been given with love. The original Croghan silver tea service which now sits on the sideboard at Locust Grove must have included the silver tankard as part of the original tea service. Did Ann Croghan Jesup receive the silver tankard from her family and pass it to Mary when she married? The Blair family left Silver Spring and all furnishings to daughter Lizzie who had lived there with her family for thirty years. General Jesup's family had no heirlooms to give and the Jesup sisters never knew their father's family, so many unanswered questions remain.

George Wheeler was shot and killed by another man one year after Jim died. Since neither Jim nor Wheeler had a will, all of Mary's possessions given to Jim were sold at an auction house in New York.

Following the death of Mary's two younger children, Mary and Lucy Ann visited often. Their two daughters, Violet Janin and Mary Sitgreaves, vacationed with their mothers most every year at some fashionable New England seaside location. The sisters corresponded almost daily.

Mary and Violet were the force behind organizing The Washington D.C. Colonial Dames and the Daughters of The American Revolution. All the Jesup sisters were very active in these patriotic organizations in honoring their Colonial ancestors. Some of the first meetings were held in the Blair House. Mary was president of the Washington D.C. Colonial Dames chapter and she and Violet are responsible for collecting many Revolutionary War relics for Dumbarton House.

Lucy Ann Jesup Sitgreaves died, 1912. The next two years seemed to be the loneliest in Mary's life. She died peacefully on June 4, 1914 and was buried in the mausoleum where James Blair was buried. On December 26, 1912, General Thomas and Ann Croghan Jesup's remains were placed in the Arlington Cemetery.

PART V

The Croghan Heirs
of Mammoth Cave

CHAPTER ONE

The Croghan Heirs
of Mammoth Cave

In the 1840s Dr. John Croghan of Louisville, Kentucky had a visionary dream of an extraordinary piece of property which he had acquired in Edmonson County, Kentucky and was in the process of developing it as a sightseeing tourist attraction.

If Dr. Croghan could return today and see what has happened to his dream, he would see a National Park which has grown around the Cave as one of the most visited places in North America. Tour buses, school buses, cars and other vehicles wait for a place in the large parking areas. This work of nature became a National Park in 1941, but was not formally dedicated until September 18, 1946.[1]

Mammoth Cave has been called one of the seven "wonders of the world".[2] Inside the cave are corridors on five different levels, with formations of stalactites and stalagmites and a river. Some descriptive words used by writers are: "spectacular," "magnificent," nature has shown her handiwork to tell the story of Mammoth Cave. Different passages, or "rooms," have interesting names. "The Gothic Avenue is so named because of resemblance to the old cathedrals of Europe. The Cleveland Room is a room of beautiful formations. Some formations seem to be crystallized diamonds. Sparkling substances, resembling stars, is appropriately named the 'Star Chamber'. The immense distance to the ceiling, the visitor inside fancies he is standing under the canopy of heaven."[3] The beauty and wonder of this cave is religious in every respect.

Western Kentucky was an area that Doctor Croghan knew very well, owing to the fact that his father, Major William Croghan, a Revolutionary War Veteran, had surveyed and claimed thousands of acres in this area. The Major had developed iron works near Mammoth Cave and had established the town of Smithland where the Cumberland empties into the Ohio River.

Commercial trade on the Ohio and Mississippi Rivers was also, a financial success for Major Croghan.

Interest in the Cave increased for Dr. John Croghan of Locust Grove, Louisville, Kentucky, while traveling in Europe in 1833. He was surprised that interest in the Mammoth Cave discoveries were great news and sought after by newspapers. He began to see this natural wonder in a different light. Dr. Croghan visited Mammoth Cave in 1839 and listened to Franklin Gorin tell about the increase in business, but evidently a larger inn was needed.[4] The Croghans had business sense and Dr. Croghan recognized this great piece of property had potential for drawing greater worldwide attention. He didn't let this opportunity pass which was bound to be a good business investment.

For $10,000, Dr. Croghan purchased the Mammoth Cave and three slaves who served as guides and were part of the sales agreement.[5] One slave was Steven Bishop, a famous tour guide who was a great asset to the business. From many accounts, no one was more responsible for exploring, discovering and opening other chambers in the cave than Bishop. By the time, Croghan made his purchase, Steven Bishop's fame had grown and newspaper editors were anxious to write about each new discovery in the Cave. He was the first to cross the daring and dangerous pit of the Bottomless Pit on cedar-pole ladders. "Bishop was a self-taught man, his intelligent face is assured and tranquil, and his manners particularly quiet. He talks to charming ladies with the air of a man who is accustomed to their good will and attentive listening. On Echo River, he sang with a full melodious voice, a three-tone sequence of notes which came back a splendid chord."[6]

Two more slaves were hired and trained by Steven. Very soon, all were kept busy as guides on the tourist trails of the Mammoth Cave.

One of the first business ventures at the Cave was to construct a new road system. Dr. Croghan built a network bringing the outside world in. Travelers had been stopping overnight at a place called Bells Tavern which was not on one of the new roads. A new four horse stage route led to the Mammoth Cave.[7]

Dr. Croghan was so impressed with Bishop, including his ability to draw maps. He asked Bishop to sketch a map of what he had found. The next winter Croghan took Bishop to his home, Locust Grove in Louisville. He was there for two weeks and drew the map in ink with such likeness that Croghan had copies made for use at the Cave. Bishop was given full credit for his cartography. Croghan looked upon them as things of beauty and Bishop's reputation grew.[8]

At forty-five years of age, the doctor, was convinced that the climate in the Cave would be healing to consumptive (tuberculosis) patients and the pure air would be very beneficial. In addition to a tourist resort, a sanitarium was built inside the cave and was referred to as an "invalid's village." The patients lived in private cabins. Needless to say, the damp conditions in the Cave, coupled with the smoke from cooking and heating, hastened their deaths.[9]

Dr. Croghan spent his last ten years building the Cave into a famous tourist attraction. He died in 1849, leaving Mammoth Cave to his nine nieces and nephews. During the next one hundred years, this piece of property became so valuable that many would fight for control of the Cave. In the end, the courts resolved the question of ownership.

Visitors to Mammoth Cave are unaware of the difficult years which took place between nine Croghan heirs (nieces and nephews of Dr. Croghan) as a result of each having equal power. The family struggles and infighting was bound to occur among nine heirs with no controlling power at the top. This situation continued for many years, in spite of good intentions.

On February 5, 1849, Dr. Croghan's will was probated which placed the tract of land called The Mammoth Cave and Salts Cave in trust for his nine nephews and nieces. These nine heirs were grandchildren of Major William and Lucy Clark Croghan.

Eight Croghan children were born and raised at Locust Grove, including Dr. Croghan. At the time the Doctor wrote his will, there were ten grandchildren, nine of which became heirs of the Mammoth Cave property. Three of the Mammoth Cave heirs were the children of second son, Colonel George Croghan and six were the children of daughter Ann Croghan Jesup. George and Ann's other brother, William Croghan, Jr., left his only child, Mary Croghan Schenley a fortune which made her a multimillionaire. She lived in England and one would assume that Dr. Croghan felt this niece was well provided for and excluded from the will.

Nine Mammoth Cave Heirs

George and Serena Croghan's Children

1. Mary Angelica Croghan (Wyatt)
2. St. George Croghan
3. Serena Croghan (Rodgers) Last Heir Died 1926

Ann Croghan and Thomas Jesup's Children

4. Lucy Ann Jesup (Sitgreaves)
5. Mary Serena Jesup (Blair) Financial Investment saved
   Mammoth Cave
6. Jane Findlay Jesup (Nicholson)
7. William Croghan Jesup
8. Charles Jesup
9. Julia Clark Jesup

According to the terms of the will:

1. The said lands and buildings were to be rented (except the Cave)
   for terms of five years.

2. Trustees were to appoint an agent to run the property, hire
   guides and servants, and provide for anything necessary to
   exhibit the Cave to visitors. The agent was to keep an account
   of all expenses.

3. Money received from visitors after expenses, was to be divided
   among the heirs.

4. When any of the nine nephews or nieces died, their 1/9th portion would go to their heirs.

5. When the last of the original nine heirs died, the trustees were to
   sell the lands and Cave, giving the proceeds to the remaining
   heirs.[10]

Dr. Croghan's intent was to provide a source of income for his nephews
and nieces. However, the first thirty five years after his death were years of
conflict. At times, Mammoth Cave seemed to be a burdensome white ele-
phant. The first Cave managers were agents who brought nothing but disappointment to the heirs and the property declined in value.

Judge Joseph Underwood, a cousin of Dr. Croghan, was first trustee of Mammoth Cave and served from 1849 to 1870. The following is an example of management reports sent while Underwood was in charge of the Cave.

> Report for 1867: Mr. Proctor, the lessee, had agreed to pay, in improvements, a great deal more, (Underwood's assessment), than the hotel and farms were worth. Underwood regards $20,000, (five year lease), was too much because the receipts would never pay rental of the property after expenses.

$5,761.75	Total amount of receipts for 1867
<u>1,914.25</u>	Guides
3,847.50	Balance
<u>3,100.00</u>	Underwood's pay
747.50	Balance
<u>494.75</u>	Proctor (lessee) Paid to Underwood
$ 252.75	Balance at the beginning of 1868

An increase in profits was made from 1866 to 1867, resulting from an arrangement made with the railroad company for group tours. Underwood offered to relinquish his trusteeship. He wrote that the only options open for the heirs were:

> Attempt to sell property under a decree of the Chancellor

> Invest proceeds in stocks until the last heir (Julia Clark Jesup) has reached 21 Select some young man to take Underwood's place.

> It requires having a great deal of means at one's command to operate successfully. Men possessing these means are not likely to become renters or hotel keepers.

> Improvements at the Cave are of wooden structures constantly needing repair.

> If nothing can be done, but go on in the old way, I will endeavor to execute the trust according to the will of my relation, Dr. Croghan and leave consequence after my death to take care of themselves.[11]

Underwood resigned in 1870.

In the 1870s, the Cave continued to lose money. As a result, Christopher Wyatt (husband of Angelica Croghan) had a plan to petition the courts for an order to break the will and sell the Cave. Much time and effort went into making this sale a reality.

This required Wyatt to visit the Cave and while there, he made a trip to Louisville, Kentucky and spent some money. He received the opinion that it was possible to break the will, but the money from the sale of the estate, would have to be invested in Kentucky, at the will of the Court. This is when Wyatt went into action to gain Power of Attorney from all the heirs. Since they lived far apart from each other, it would be convenient for him to sign and put the trust in his name for the heirs. Wyatt wrote that he already had Power of Attorney to sign an agreement for "myself [wife, Angelica Croghan] and Mrs. Rodgers [Serena Croghan] and the children of the late St. George Croghan."[12] He held power of attorney for Serena Livingston Croghan and acted as financial agent for her which began before her first voyage to California.[13]

In order to make a sale, it was necessary for all adult heirs to sue the secondary minor heirs. February 1876, Jesup Blair received a letter from Wyatt in which he was urging the three trustees, Jesup Blair, Nicholas and Sitgreaves, who lived in Washington D.C., to sign the agreement giving Wyatt authority to sell the Cave. "Please act as quickly as you can in confirming with others, & sign the agreement, & return it to me; when I will also sign it, & put it in the hands of Mr. Ogden Clark for immediate action. Yours truly, Chrs. B. Wyatt."[14]

The Jesup sisters would not allow the Washington D.C. trustees to sign the agreement, so they were sued along with the under age heirs.

At that time, George Croghan III, a twenty-six year old grandson of Colonel George Croghan, was in Kentucky representing the absentee heirs of California in bringing a lawsuit to sell the Cave for $110,000.[15] George made affidavit that all the plaintiffs, except himself, were absent from Edmonston Co. and then alleges as to the addressing of the other parties to the suit.

Office of BYRON RENFRO
ATTORNEY AT LAW
Louisville, Ky. ...1870

Edmondson Circuit Court

Mary A. Wyatt ....**Plaintiff**

Petition in Equity

Mary A. Wyatt
Wm E. Wyatt
Chris. A. Wyatt
Mary L. Wyatt
Francis V. Allen and Henry R. Allen
   her husband
Serena E.L. Rogers and Augustus Rogers,
   her husband

} **Plaintiffs**

Against

Cornelia. L. Rogers ....**Defendant**

Cornelia. L. Rogers
Mont. Rogers
Mary Rogers
Aug. "
Henry "
Nannie "
Grace "
Robert ")
Lucy Ann Sitgreaves and Lorenzo Sitgreaves
   her husband; Mary J. Sitgreaves;
   Mary J. Blair; Jesup Blair
Jane F. Nicholson and Aug. S. Nicholson her
   husband:
Julia Jesup:
Violet Janin and Albert C. Janin, her husband;
Lucy J. Wheeler and George Wheeler, her husband

} **Defendants**

This attempt to sell the Mammoth Cave failed. Dr. John Croghan's will was upheld. As a result, one of Underwood's suggested options was no longer a possibility.

In 1876, Wyatt was once more encouraged that a sale may be affected with buyers from England. The buyers had visited the Cave and proposed to go (the elder of the firm) to England, forthwith to try to bring about a sale. Wyatt was much pleased with the energy they displayed. This sale did not occur and soon after, Wyatt warns, "If the Cave fees cannot be held in reserve, we must advance money to sustain it. I am not prepared either to advance money, or to incur debt for conducting the business of the Cave."[16]

None of the original heirs lived in Kentucky and after 1861, only six female heirs were living. Four Jesup sisters lived in Washington, D.C. and two Croghan sisters lived in California. Various schemes to improve the amount earned by the Cave went awry. For years the female heirs were not consulted about Cave matters. The four husbands: Colonel Sitgreaves, Major Augustus Nicholson, Christopher Wyatt, Augustus Rodgers and one son, Jesup Blair, acted as trustees. Financial reports were sent directly to these men. No one was happy with the situation at the Cave, even the trustees.

Wyatt had urged the heirs repeatedly to visit the Cave. He warned "that it is simply impossible for anyone to derive or decide intelligently as to what we can do, ought to do, or may expediently do, with the Mammoth Cave Estate, until the individual has gone upon the ground in person, & tarried long enough to investigate & understand the many circumstances peculiar to the cave, & effecting in one way or another, the formality of any measure."[17]

In reply to one of Jesup Blair's letters, Wyatt said, "My suggestion to you would be, inasmuch as you are a man of abundant license and possibilities, you should first of all visit the M.C. and pass a week in investigating the whole condition then, give us a report. It is your turn. If I had not acted on three different occasions the Cave would have been lost to all the heirs"[18]

In October, 1876, Wyatt's patience had worn thin with the Jesup women. They had begun to speak up and wanted changes. He told Jesup Blair, "Nix to the wishes of Mrs. Blair and Miss Jesup. Any meaning which I can attach to those words is inconsistent with the fact that Major Nicholson was requested to act, & did act, as principal & chief of the Trustees, because he might therefore easily confer together with him. Without action of a majority of the trustees, I consider myself or any individual member of the board, incompetent to direct any change, or to give orders in contradiction, of the policy adopted by the whole board, after careful consultation of the wishes of all the women."[19]

Christopher Wyatt died sometime before 1881. Conditions at the Cave had deteriorated with an agent, Mr. Miller, who failed to do a satisfactory job.

George Croghan III was ten years old when his father St. George Croghan was killed in the Civil War, 1861. In 1880, the Jesup sisters sent George Croghan III and Mary Jesup Blair's son-in-law George Wheeler to the Cave and upon showing Miller his power of attorney, Wheeler proceeded to examine the books. He could not believe the condition of books and vouchers and declared it was useless trying to ascertain anything he wished to be informed upon. Mr. Miller kept no pay vouchers and when Wheeler called his attention to this, it was confusion and even worse, confounded with obstacles and hindrances thrown in his way the whole time he was there. Wheeler departed after two weeks, leaving a letter for George to send to Mary Jesup showing how much in debt the heirs were.[20]

Wheeler's letter to Mary included a request to make a copy for Major Nicholson, Washington D.C., showing the receipts and expenditures for the two weeks he was there. The copy was to be certified by Mary for authorizing Mr. George Croghan to assist her in the clerical work of preparing the same. Mr. Miller sent his own copy to Major Nicholson, but no one was allowed to read it.[21]

George was met with deplorable treatment from those at Mammoth Cave when he started making changes. He discovered the managers' families had moved in, free board and meals. Some were hauling supplies away in great amounts and selling them. George Croghan took measures to change this and daily sent reports to the heirs, informing them of his discoveries and steps he was taking to correct the conditions, but his efforts only lasted for a short time. Orders and instructions arrived at the Cave from those in power, informing the employees that no one was to take orders from George Croghan and immediately he was informed of this by the employees. They would not respect his actions in any form and most of all, he was not allowed to see the bookkeeping. At the instruction of Nicholson and others, two gentlemen (cousins) arrived to assure George Croghan left the premises. But, before the two arrived, George had skillfully gained access to the bookkeeping room and time was on his side long enough for him to learn that $90,000 had been embezzled. At gun point, George was warned never to reveal what he had discovered or else.[22]

The removal of George was not ignored by Jesups and they immediately took control of management at Mammoth Cave by paying the indebtedness and bought all buildings and operations of the cave complex. The youngest Jesup sister Julia left the busy whirl of Washington D.C. society and arrived at Mammoth Cave as resident agent Trustee. There were no doubts as to who was in charge thereafter.

A new manager Mr. Kleits had been hired and he had his own method of keeping the books. Augustus Rodgers, in California, wrote to Augustus Nicholson that "duplicate receipts should at least be taken and the duplicates forwarded monthly." He felt this "loose way of doing business is an encouragement to dishonesty and it is poor business management with one man in the hotel, another in the Cave, and another in charge of transportation and each working as they have often done, against all interests but his own." He believed that the steam [boat] had made it easy to find richer opportunities as westward expansion grew.[23]

Julia spent five years acting as resident Agent Trustee at the Cave. Augustus Rodgers sent long letters of advice to her. "Excuse me if I am unnecessarily explicit. I find such a large majority of women uninformed in business *details*, I am not certain of you."[24] Before Julia resigned in 1885, Rodgers must have had a change of heart because he wrote, "Through your personal attention, we now know what we never knew before that the Cave is a 'paying property'. You certainly have done far more than any agent we ever had & have demonstrated some very curious things in the way of management."[25]

Perhaps the "curious things" were in reference to the time Julia thwarted Jesse James' plan to rob Mammoth Cave of $6,000.00.

> In the late summer of 1880, while several members of the James gang were sojourning at the hospitable home of Father Hite in Logan Co., and were receiving their mail through the kindness of Miss Nannie Mimms, the boys were going out in twos and threes into neighboring counties to steal horses or pick up a bit of pin money by robbery. Once Jesse and two others set out for Edmonson county to rob the stage between Mammoth Cave and its railroad station, Cave City, but it rained too hard, and they turned back. Later, about the first of September, Jesse and Bill

Ryan rode up there and accomplished the job, holding up two coaches, one going each way, within an hour of each other.

Miss Jesup, spinster, of New York, one of the original Croghan heirs who then owned the Cave, happened to be at the hotel at the time, and had prepared, so we are told, about $6,000 in cash, the receipts for several weeks past, for shipment to a bank. It would seem as if the plan to rob the stage was based on inside information, but such was not the case. It was purely a coincidence.

However, Miss Jesup had a real hunch. Probably she had been reading in the newspapers of the holdups in the counties below. Anyhow, she was afraid to entrust that money to the stagecoach, so she sent it in six packages of about a thousand dollars each, by Jim Brown, a Negro servant, on horseback, treading woodland trails, cutting across gulches, and rocky ridges to Cave City, where the money was safely placed in the express company's hands.

With no express treasure aboard, the only thing for the bandits to do was to stand the passengers in line and go through their pockets.[26]

George Croghan III was a passenger when the stagecoach was held up. The bandits took watches, very little money off the passengers. They did take a bottle of whiskey from George. One can not help being amused when reading George Croghan's letters while in Kentucky. He declared that anything worth talking about happened "... befo the wah."

Jesup Sisters

Mary Jesup Blair's son, Jesup, died April 1, 1902. Grief sometimes turns into energetic anger. This could have been the case with Mary. At that time, the Jesup sisters were beginning to exercise their authority as "rightful heirs." They held two-thirds ownership in Mammoth Cave and the Croghan sisters held one-third. At the turn of the century, Mary Blair, Lucy Sitgreaves and Julia Clark were against the nomination of their sister Jane's husband, Augustus Nicholson, as a Mammoth Cave Trustee.

The strains on the Jesup sisters began to show. Lucy Jesup Sitgreave's daughter, Mary, made the decision to remove her mother from the friction in Washington D.C. to Boston, Massachusetts. Lucy could not bring herself to appear in court, bringing charges against her sister Jane Jesup Nicholson's husband. It seems Nicholson had committed some unscrupulous act of fraud in the business at the Cave and Mary Jesup Blair never forgave him. Nicholson wasn't the only one. He just wasn't as clever as others.

The youngest Jesup sister, Julia, had lived for a short time in Europe after the Civil War and evidently planned to return. She moved away from Washington D.C. to Europe in the late 1880s and it appears, she never returned for a visit. Close family ties existed among the transplanted Clark descendants living in Europe. Julia lived in Italy where some of Jonathan Clark's descendants had made their home. Long visits developed between Julia, Cousin Mary Schenley and her family who lived in England.

Mary Jesup Blair was left to fight alone for the survival of family interests at Mammoth Cave. It wasn't easy and there were times in which she must have felt very much alone. She was also having trouble with her daughter Lucy's husband, George Wheeler.

Mary's son-in-law, Albert Janin, who came from a prominent family of lawyers in New Orleans had reached the age of retirement. He was needed as resident manager at the Cave and was to represent the Jesup side. Strangely enough, Nicholson appealed to Rodgers in California, who, in turn, sided with Nicholson. These partners, who had held power for years, but were not "rightful heirs," fought back.

Albert Janin soon discovered that money had been missing over the years and unaccounted for. Nicholson was one of those involved. Janin was going to court over what could be called nothing other than "theft."[27] This caused painful friction between the Jesup sisters. Even though Lucy Sitgreaves and Julia had left the whole disagreeable Cave business to their sister Mary, they always remained in contact with each other and Mary and Lutie visited each other regularly.

Augustus Rodgers protested Janin's presence at the Cave. Janin responded with a clear understanding, "I haven't the least feeling of personal enmity towards our co-trustee, Nicholson. The fact is, I care nothing about him and I am so constituted that I do not waste time upon or burden my emotional nature with, personal hatreds and resentments. But, I regard N. as the evil genius of the Mammoth Cave and I feel it to be my duty to antagonize him what I believe to be his secret persistent aim and determination to obtain, again, control of the M.C. business."[28]

One of several "clear understandings" made by Janin was "that in the future, the power and decisions were to be made by the *direct heirs* with checks made in their name!!"[29]

Mary Blair and Lucy Sitgreaves had been critical of Rodgers after he had very strongly expressed his displeasure about the power struggle. He told them, "I regret that my judgment in Cave affairs should be adverse to yours, life long friends of mine, as both of you have been."[30] He deplored family misunderstandings and in the Cave matter, he realized the weakness and disadvantage upon the business with "a house divided against itself."[31] Rodgers was adamant in his belief that managers must always be non-resident trustees and have control over the property and their decision as to who shall be the resident manager and be accepted by all.

Rodgers deplored the "unpleasantness" caused by Agent Nelson, who was accused of stealing from Cave property by use of a special brand for Cave merchandise without authority. Throughout 1905, letters flew back and forth among the heirs regarding the date of Miller's resignation and impending law suits against him by Janin. Augustus Rodgers was greatly against such suits, especially as they concerned the California heirs. He noticed that the Cave Agents in the past few years seemed to make the same amount as the heirs.

Augustus Rodgers never took too kindly to the news that Janin was resident Trustee, and in 1905 he was still writing sarcastic letters. He informed Janin, "In the future, the checks for Mrs. Rodgers, Mrs. Browne and Miss Horner could be sent to them and acknowledged by them since power and decisions now, were to be made only by the direct heirs. Both Mrs. Rodgers and Miss Horner are both greatly dependent on Mammoth Cave returns. Both of them are 'heirs in fee' as you would express it. I am very sorry that the ordinary courtesies of business should be omitted in the failure to acknowledge receipts. Do you say 'Please acknowledge receipt in sending reports and drafts?" He goes on to advise against coal oil lamps which would eventually mar the sparkling beauty of the Cave. A purchasing agent would electrify the Cave. Rodgers likes the idea of waking up the weary, on return, with simple refreshment.[31]

Janin managed to make improvements to the Cave and the monthly checks sent to the heirs increased. Even during World War I in 1918, when business was down, Serena (Tinie) Rodgers' check for one month was $468.00. Dividends in winter months were always lower, but overall, there was a noticeable increase in monthly dividends after Janin took charge. Perhaps, this happened as a result of Janin's creative ideas and advertising. The lifestyle for Croghan heirs began to change. Travels back East became quite frequent and descendants of George, William and Ann Croghan learned to know each other once more. On voyages abroad, they became acquainted with their Schenley cousins.

After Mary Jesup Blair's death in 1914, Lucy Croghan Browne was sharing some sweet memories of "cousin Mary" with Violet. "I prize the Prayer Book bearing the name of 'Lucy Clark' which your dear Mother gave me & remember the old pew in St. Johns Church where we sat together remembered the loving kindness of Mary Jesup Blair giving her Lucy Clark Croghan's Prayer Book."[32] Correspondence between Colonel George Croghan's and Ann Croghan Jesup's descendants was warm and loving.

Mary Jesup Blair's son-in-law took charge of the Cave in 1905. He had not been successful in his pursuits to become wealthy while living in New Orleans for many years. Janin had made bad investments losing large sums of borrowed money from Mary, his mother-in-law. He was the hierarchy of the Cave now, and would prove himself by showing the world how things were done in New Orleans by bringing "New Orleans Style" to the guests in Kentucky. To everyone's surprise, he made a success of the Cave's financial affairs.

H.C. Hovey in 1909 described the hotel as follows:

> A bugle flourish would herald the arrival of passengers and bring around the coach a throng of guests expecting to see friends or curious to see strangers, together with a sizable group of Negro servants ready to offer their services and take care of the luggage. The hotel register showed from two to three thousand visitors a year. Many came from the north and a few from various parts of Europe. The majority however were from Louisville, Nashville, Memphis, New Orleans and other cities of the South. Loitering along the long colonnade in the evening, guests would look between tall white pillars through the noble grove of aged oaks and across the bluegrass lawn. At 11 P.M. the band left the ball room for the veranda and according to their custom gave the signal for retiring by playing *Home, Sweet Home*. The next morning at six o'clock the same musicians awoke the guests by playing *Dixie*.[33]

In January 1912, Lucy Browne in California was desperately seeking information from Janin about her brother George Croghan III, upon learning of his death in December. Even though he lived somewhere in California, relations were strained between him and his two sisters. They knew nothing of the cause of his death. Lucy and her sister, Elizabeth (Lily) Kennedy, wished to place a stone on his grave but did not know the exact location. After four generations of Locust Grove Croghans, the last to carry on the name *Croghan* ended with his death. He had never married and had no descendants. Lucy Browne asked Albert Janin about George's membership in the Order of Cincinnati, as her son, Spencer Browne, would succeed him in that order as the next in line.

There was great unrest in Kentucky at this time, concerning what was happening to the natural beauty of the Cave area. Cave onyx was being stripped for sale to tourists. Janin was accused of cutting virgin forests. (This could have been in reference to the YMCA Camp and new roads he built in 1921). Unscrupulous people were attempting to open new entrances to the Cave outside the Croghan property lines. In 1911, Lucy Browne made reference to a bill being brought before Congress in Washington, D.C. pertaining to purchasing Mammoth Cave. A Mammoth Cave National Park Association was formed by a group of Kentuckians to protect the area.[34]

Cornelia Rodgers Nokes and Lucy Croghan Browne were always apprehensive about Cave problems and Janin, alone, had to deal with them. Janin received a letter from Cornelia,

> Just as my Mother finished her letter to you, a man came who represented himself as a friend of Benj. F. Nasser [Natcher?], Kentucky, who wished him to find correct names and addresses of Mammoth Cave heirs in California, that he understood the principal life owner, in California, had died. My brother told him that he had been misinformed, that our mother was in excellent health. I am writing you this, as perhaps these inquiries are something you may know about.[35]

Cornelia wrote to Janin again,

> The telegram I forwarded to you was from a Kentucky paper to a newspaperman here, asking if Mrs. S. L. Rodgers was still alive. I can not help feeling that this so called 'Cave Development Company' is waiting for her life to cease to try and get possession of the whole property.[36]

Again, in 1924,

> I am much astonished and indignant that you should have been given the impression that in any way, I have had any communication with anyone in Louisville. The enclosed telegram to my Mother, I answered at Mr. Wyatt Allen's advice and on the back, you will see my reply jotted down as I sat at the telegram. 'Mrs. Rodgers not sufficiently informed refer you to Judge Janin.' That was all, I did not even sign my name. Please believe that you can entirely trust the heirs in California to rely upon what you are doing. I am very sorry that you should have allowed yourself to believe this of me.

(Notation written by Janin "I don't know what she is referring to. Know nothing about it, A.C.J.)"[37]

In California, the Rodgers and Browne families lived near each other and were very close. A special bond had grown between the California cousins and Violet Blair Janin in D.C. as a result of correspondence and Lucy Browne's trips back East. After 1922, Violet and Mary Sitgreaves were the only surviving members of the Jesup family. Lucy Browne (California) continued to send letters to Violet and in one, gives information about "Aunt Tinie" (Serena Croghan Rodgers) "… very deaf and nearly blind, she has not much to live for, has she: But, she has no suffering and that is a blessing. Her children are very devoted and keep her very comfortable. She is a perfect little saint, so patient, but longing to go."[38]

Tinie's health began to fail two years before she died, but Cornelia kept family members informed about her mother's well-being. "Mother lives more in the past than in the present and remembers so the old times and has continued her girlhood days. She is very well though, remarkably so, goes out to church and to walk or drive. We read to her a great deal."[39]

There was a never-ending fear in Cornelia's life that the Cave dividends would be taken away. A letter in April 1924, sounds as if she is in a panic concerning a letter from Mr. Van Wyck (Lucy Browne's son-in-law) with his unusual simile of the "Goose and the Golden Egg" as pertaining to Cave dividends. Two of the heirs are almost dependent upon the income. They have had to supplement Miss Horner's income each month since her dividends from the Cave were not sufficient for her needs. Cornelia felt that her mother and herself couldn't support Sophia in case the dividends should stop. Nor could Sophia's aunts, Mrs. Browne and Mrs. Kennedy.[40]

About this time, the L&N Railroad was playing an active part in promoting the establishment of a Mammoth Cave National Park Association. The L&N railroad owned 3,300 acres of land, caves and cave rights. A proposition was made by the railroad company to the citizens of Bowling Green, Kentucky to support the formation of an association. May 1924, representatives of the railroad and local citizens succeeded in forming the Mammoth Cave National Park Committee.[41]

The Croghan Heirs were very concerned at that point. Spencer Browne's name comes up often in his mother's (Lucy) letters. He was in his last year of college in 1905 and studying mining engineering. After a visit back East in 1911, Lucy wrote, "Spence and friend have just started on his way back in his motorcar, coming by Southern Highway which will take some weeks." In 1924, this young man was living in New York, had visited Mammoth Cave and gave a very important "wake-up" call to three of the remaining secondary heirs: his mother, Lucy Browne, Cornelia Nokes, living in California and Mary Sitgreaves, living in Boston.[42]

From Santa Fe, New Mexico, Spencer wrote to Mary Sitgreaves in Boston. Spencer was very disturbed about finding Violet Blair Janin, bravely maintaining the position in place of her stricken husband, who had been controlling the Cave property. He felt the situation was very serious, and so full of danger to the interests of all the Cave owners, that it was advisable to explain the course of action that he had urged Violet to adopt.

At that time, there were only two trustees of the Cave property who were nominated by the owners. Mr. Janin, the Managing Trustee, had given many years of close attention to Cave affairs, but Mr. Janin was eighty years old, ill, and no longer competent to act. Violet admits that he is physically feeble, in danger of a stroke, his mind is affected at times and there is no hope of ever recovering his health. The entire control of the Cave was about to be passed into unfriendly hands.

Tinie Rodgers, the remaining heir, had just celebrated her ninetieth birthday and was feeble, sometimes had sinking spells. Six months after Tinie's death, the Cave would have to be sold.

Spencer tried to stress how important it was to have the sale of the Cave conducted by trustees friendly to the owners, so that every opportunity should be afforded to obtain a full and fair price. It would be disastrous for some of the owners, if the Cave should be sold for a small portion of its value.

The three remaining secondary heirs are reminded that Violet has had to leave her comfortable home and her congenial pursuits with friends back in Washington D.C. She has had to spend many months in uncomfortable surroundings at the Cave, but will have to stay as long as the present situation endures. Spencer was so impressed with Violet's courage and fighting spirit, that he thought her strength was worthy of their great ancestors. She was bearing the entire responsibility of the Cave management on her shoulders.

Spencer said this was very unfair to Violet and to all the owners. There was danger of losing control of the Cave property. He urged the owners to pay someone a part of the Cave income, rather than lose it all through failure to take adequate legal precautions.

"Judge William H. Hunt, has recommended the affairs be placed in the hands of a firm of lawyers of national reputation. The nonresidents of Kentucky should get the case transferred to a Federal court, rather than be influenced by local politics or intrigue."[43] Spencer urged the owners to do this before Janin became legally disqualified to act and wrote the following;

> My own recommendation endorses this, and goes further. I have advised Mrs. Janin to take advantage immediately of what authority remains in her husband's hands, and before he might become altogether disqualified physically or legally, to have him appoint such a firm of lawyers to act for all the Cave owners. then the lawyers should be appointed to the vacant trusteeships, and the duty imposed upon them to appoint and control a managing agent, and the entire responsibility of the Cave management be thus placed on their shoulders before Mr. Janin becomes legally disqualified to act.
>
> I am not recommending any particular firm of lawyers, but believe they should be of the highest standing for capability and integrity, and should come, not from Kentucky, but from New York, Washington or Boston where most of the Cave owners live.
>
> Judge Hunt had presided over cases of the highest degree of importance, such as the suit to dissolve the Steel Corporation. I am writing him asking him to recommend several Eastern firms, and will forward his recommendations to you and to Mrs. Janin.[44]

This recommendation was interpreted by the Rodgers family (and possibly Janin), that Spencer had a hidden agenda which would be in Spencer's favor, by appointing him or someone from his firm. The simile made by Wyatt Allen about "the goose and the golden egg" was in reference to this supposition that there was a hidden motive behind the suggestion to hire lawyers of highest standing and not from Kentucky. Wyatt Allen was wrong and the California heirs should have listened to Spencer, who felt that local courts were run by local politicians.

> Again, let me warn you that the only legal control of Cave affairs is now in the hands of Mr. Janin whose health and mind are waning fast. If the owners do not soon find strong shoulders to relieve Mrs. Janin and to bear the responsibility of administration, they may lose the entire control and income from the Cave.[45]

On May 12, 1924 a document signed by Serena Croghan Rodgers, authorizing Cornelia Nokes, Violet Janin and Mary Sitgreaves as trustees was delivered to Janin at the Cave.

True to Spencer's fears, Edmonson County Circuit Court would not acknowledge the designated family appointed trustees. They sued for the heirs, but lost.

Later in September, Spencer Browne wrote to his cousin, Violet, that a firm had been recommended to him, but by December, Violet had found a lawyer herself, Mr. Marshall Bullitt. Violet suggested in April, 1925 that a syndicate of the beneficiaries be formed and that Spencer head such a group. He felt that it would require "more time and attention than I personally could spare to obtain the frequent approval of thirty-odd interested parties." He hoped that Violet would drop the "details of litigation" and direct her keen mind to the "demonstrating the priceless value of the Cave in advance of its sale."[46] Evidently, Spencer knew the Cave was already lost by that time.

At the same time, in 1925, Congressman Thomas had pushed for a national park for Mammoth Cave but had no hope of it being approved in the current congress. The administration of the President offered no encouragement for the project which was rejected from the President on down. There was no certainty that the Secretary of the Interior would recommend it, even though enough land had been acquired and the government would not have to spend any money for the project.

Spencer Browne realized the die was cast when he declined to accept Violet's request for him to head a syndicate of the beneficiaries to be formed. Marshall Bullitt was related to the Clarks, but he was still one of "those" Kentuckians. By this time, even if Marshall Bullitt had been a respected powerful attorney from New York or Washington D.C., there was no chance, then or ever, for a syndicate of beneficiaries achieving success against the following powers of Kentucky:

J.L. Harman, John B. Rodes	Representative M.H. Thatcher
R.C.P. Thomas, Mrs. B.W. Bayless	Mrs. Thatcher
Eugene Stuart, Alex E. Johnson Jr.	Representative R.Y. Thomas,
Milton Smith, Robert J. Ball	Alben Barkley
J. Graham Brown	Lee Lemar Robinson

The entire delegation of the above had gone to the Interior Department in Washington D.C. Hearings were made to Congress, were passed and signed by President Coolidge on *May 25, 1926.* The lands would be secure by public and private donation.[47]

Last Heir of Mammoth Cave Dies 28 August 1926

This news of Serena Croghan Rodgers' death solved some problems for the secondary heirs, but there would be new problems to face.

The Mammoth Cave Association and the M.C. Commission now had the task of buying up enough surrounding land to establish a national park. The fund-raising started. It seems that when Congress approved the act for a national park, the same legislature required that The Mammoth Cave Park Association was to deliver 70,618 acres, before the national park would be established. The heirs refused an offer of $300,000. It became apparent that a declaration of eminent domain would be necessary to acquire enough land. Eminent Domain meant the right of the government to appropriate private property for public use, usually with compensation to the owner.

The private Mammoth Cave Park Association could not exercise the right of eminent domain, so in 1928, the Kentucky Legislature formed a Mammoth Cave Park Commission. The Governor was to appoint the members of the commission and the chairman would be the Governor. The Commission condemned the entire cave and in 1930, the property was acquired by the state of Kentucky.

Soon after Janin's death, Violet Blair Janin sold her family's interest on December 31, 1928, for $446,000 to the United States government.[49]

After Violet Janin sold her share of the Mammoth Cave in 1928, it seems the following has been accepted as fact, as to what took place between The Mammoth Cave National Park Association and the secondary heirs.

> The succeeding heirs pooled their fractional interests in the estate comprising Mammoth Cave. and 1610 acres of land—and turned it over to a trust company in Louisville that issued certificates in return. It is commonly thought that the Mammoth Cave National Park Association bought these participation certificates one by one from the various secondary heirs until control of Mammoth Cave was acquired.[49]

The federal government established the cave as the Mammoth Cave National Park on July 1, 1941, but due to World War II, the park wasn't formally dedicated until September 18, 1946.

In the past, Violet's husband, Albert, had failed to gain success in his pursuits to become wealthy. For all his debts owed to Mary Jesup Blair, he paid his dues, not with money but by coming to her rescue when she needed him at the Cave. When Albert Janin took charge of managing the Cave, very soon the family heirs were receiving larger dividends. They recognized Janin's management and advertising were responsible for this increase in dividends and never failed to heap praise and credit on him.

Mary Jesup Blair

Had it not been for the persistence, strength and stand taken by Mary against unpleasant insurmountable dealings (which everyone else shrank from), the story would have been different. She alone was left to save the cave.

If anyone inherited the Clark and Croghan genes for greatness, it was *Mary Serena Jesup Blair.* She was the one that stepped into the role as "parent" when the Jesup brothers and sisters lost their mother, Ann Croghan Jesup in 1846. Julia was only six years old at the time and it soon became necessary for someone to take disciplinary control when Julia decided she was the ruler of the Jesup household. The oldest Jesup daughter, Lucy Ann, was a delicate beauty, who depended on someone else to take on the difficult tasks that came her way.

Janie's husband, Augustus Nicholson had to be removed as a Cave trustee and would be brought to court for unscrupulous dealings at the Cave.

Mary had to make hard and difficult decisions to accomplish the task of making the Cave a financial success. Would the state of Kentucky have put forth a strong fight to take over if she had allowed the Cave buildings and grounds to deteriorate. She had invested money making it a financial success. Afterwards, businessmen and politicians were eyeing this coveted piece of property and eventually hired lawyers to gain control. Fortunately for her, her son-in-law came just in time and under her direction, the cave was saved which provided a comfortable living for the remaining heirs.

> Of course, Dr. John Croghan's foresight in preserving the cave and its surrounding territory as a natural estate for 77 years and his own responsible stewardship of it for ten years set a pattern which led naturally to its eventual dedication in perpetuity as a facility for all the people of this country as well as those from beyond our shores who wished to visit it.[50]

The nine original heirs died, one by one.

1860, William Croghan Jesup
1861, St. George Croghan
1862, Charles Jesup
1905, Angelica Croghan Wyatt
1912, Lucy Ann Jesup Sitgreaves.
1914, June 6, Mary Serena Jesup Blair.
1922, Janie Jesup Nicholson.
1922, Julia Jesup.
1926, Serena Livingston Croghan Rodgers.
1928, Albert Janin died.—Violet sold Mammoth Cave, December 1928.
1933, January 14, Violet Blair Janin died.
1940s Mary Sitgreaves was the last Jesup Heir who died after World War II.

There are no living descendants of Ann Croghan and Thomas Jesup. Many descendants of George and Serena Croghan survive. Major Croghan and Lucy Clark raised nine children, but only one (Colonel George Croghan) has descendants in America. They still keep in touch with those at Historic Locust Grove, Inc., Louisville, Kentucky.

**Mammoth Cave: Designated a World Heritage Site—October 27, 1981
Designated a Biosphere Reserve, 1990.**

Clark and Croghan descendants at Locust Grove, fall, 2002. Photograph courtesy of Historic Locust Grove, Inc.

Sophia Horner

When great grandmother Serena Livingston Croghan died, Sophia Horner (1870–1936) was left to live in a boarding house, but visited the Rodgers household once a week for lunch. General George Rogers Clark's only surviving sword was left with this girl Sophia. She must have learned from her grandmother what this prized relic meant for America's history.

The Rodgers family only had vague memories of how the sword came into their possession. "At sometime the sword was placed on the dining room mantle by Sophia, when she came for her weekly visit." Virginia Wheaton seemed to think the sword was brought to the Rodgers house because she wasn't sure it would be secure at the boarding house where she lived. History owes a great deal to that girl.

When Sophia Horner became an adult and performing menial hospital work, the monthly checks began to arrive for the Mammoth Cave Croghan heirs. This was due to the diligence of Croghan granddaughter Mary Jesup Blair and Albert Janin's management. The heirs were obligated to acknowledge the receipt of monthly checks. When the check was more than expected, Sophia would always express her appreciation to Mammoth Cave Manager Janin for his successful management. She must have been a "girl after Janin's own heart" because if the check was $250 for that month, she was very pleased, but wanted it to be $300 the next month. Other heirs might have received $250 for the next month, but Sophia would receive $300 (probably out of Janin's own pocket).

Sophia's letters are full of pride for her Southern roots and heritage. She loved the Southern hospitality, food and music. She always wanted to visit the South where her Croghan ancestors had lived at Locust Grove. One day, she wrote to Janin about going down the street after work to attend the San Francisco Festival. She had come specifically to hear a Southern band and enjoyed their songs and music.

George and Serena Croghan's great granddaughter Sophia never married. No books have been written about Sophia Horner, but one wonders if some other person in possession of the sword, and in need of money, would have been tempted to sell the sword. Sophia died in 1936, but her General Clark sword is now displayed in a glass case at Locust Grove Museum, Louisville, Kentucky.

FAMILY CHARTS

PART I

Clark and Croghan Family Chart

I Generation

John Clark, B. 1724 King and Queen Co. VA, D. Jefferson Co, KY, 30 Jul. 1799;
Ann Rogers, B. 1734 King and Queen Co. Virginia, D. 24 Dec. 1798.

II Generation

A. **General Jonathan Clark,**
    B. August 1, 1750, Major General Virginia Militia.
    M. February 13, 1782, Sarah Hite,
    D. November 24, 1811.
B. **General George Rogers Clark**
    B November 19, 1752, Albemarle County Virginia.
    D. February 13, 1818, John Randolph called "the American Hannibal."
C. **Ann Clark**
    B. July 14, 1755
    M. October 20, 1773, Virginia, Owen Gwathmey (living in 1828
    D. October 3, 1822 at Locust Grove.
D. **Lieutenant John Clark,** 1777: Captured in Battle of Germantown.
    B. September 15, 1757
    D. November 2, 1783, Prisoner on British ship—died at Virginia home.
E. **Lieutenant Richard Clark**
    B. July 6, 1760
    D. February or Mar. 1784
F. **Lieutenant. Edmund Clark**—VIII Virginia Regiment.
    B. September 25, 1762
    D. March 11, 1815 at Mulberry Hill, Louisville, Kentucky.
G. **Lucy Clark**
    B. September. 15, 1765
    M. Major William Croghan, (1752–1822) July 14, 1789
    D. April 4, 1838
H. **Elizabeth Clark**
    B. February 11, 1768,
    M. August 14, 1787 Colonel Richard Clough Anderson at Mulberry Hill.
    D. January 15, 1795 at Soldier's Retreat.
I. **Gen. William Clark**
    B. August 1, 1770 in Caroline County Virginia.
    M. Julia Hancock, (1791–1820) Jan. 5, 1808—M. Harriet Kennerly, 1821
    D. September 1, 1838, St. Louis, MO.

J. Frances Clark
  B. January 20, 1773
  M. Dr. James O'Fallon (1766–1793) Feb. 21, 1791
  M. Charles Mynn Thruston (1766–1800) Jan. 19, 1796
  M. Dennis Fitzhugh, May 13, 1805. Fitzhugh died October 1822
  D. (Frances) Jun 19, 1825, St. Louis, MO

III Generation

Children of William Croghan and Lucy Clark

1. JOHN CROGHAN
  B. April 14, 1799, Locust Grove [Never Married]
  D. January 11, 1849, Locust Grove

2. COLONEL GEORGE CROGHAN
  B. November 15, 1791, Louisville, Kentucky.
  M. May 1, 1817, Serena Livingston of New York.
  D. January 8, 1849, New Orleans.
  Died of cholera and breathed his last on the anniversary of the Battle of New Orleans as the last gun was fired in memory of that day.
  Buried at Soldiers Monument on the Site of Fort Stephenson, Fremont, Ohio.

3. WILLIAM CROGHAN, JR
  B. January 2, 1794, Locust Grove.
  M. January 28, 1823, Mary O'Hara of Pittsburgh.
  D. September 22, 1850, Pittsburgh, PA.

4. ANN CROGHAN
  B. October 20, 1797, Locust Grove.
  M. May 1822, General Thomas Jesup
  D. April 14, 1846, Washington D.C.

5. ELIZA CROGHAN
  B. August 9, 1801, Locust Grove.
  M. September 1819, George Hancock.
  D. July 13, 1833, Locust Grove.

6. CHARLES CROGHAN
  B. June 19, 1902, [Twin] Locust Grove.
  D. October 21, 1832, Paris France.

7. NICHOLAS CROGHAN
  B. June 19, 1801. [Twin] Locust Grove
  D. July or August 1825.

8. EDMUND CROGHAN
  B. September 12, 1825, Locust Grove.
  D. August 1822.

IV Generation

Children of George and Serena Livingston Croghan

A. MARY ANGELICA WYATT
B. 1818.
D. February 1906 in New York City.
M. 1846, Reverend C. W. Wyatt.
1. Mary Livingston Croghan married Henry Newhall
    a. Alice Newhall (O'Meara): Son, Kenneth Croghan O'Meara
    b. Donald Newhall
    c. Leila Newhall, Died 1975, left jewelry valued at $250,000.
    Portrait donated to Locust Grove
2. Christopher Wyatt Married Isabelle Morris.
    a. Katherine Wyatt
    b. Frances Wyatt (1875) m. Henry F. Allen
    c. Wyatt H., m. Alysse Latham
        Alysse
        Latham Hamilton
    d. Harriett DeWitt m. Arthur E. Dodd (1906)
        Arthur, Jr.
    e. Frances, m. J. Bryant Grimwood; (1905)
        Frances
        Lucy
        Lucius H. m. Ruth Allen (1906)
        James
3. William E. Wyatt m. Jane Kirby
    a. Cornelia Wyatt m. R. Henderson
    b. Merritt
    c. Christopher B. Wyatt (1882–1930) m. Euphemia Van
    Rensselaer Waddington
        Elizabeth Wyatt m. William A. Russell.
        Jane Wyatt m. Edgar B. Ward
            Christopher Ward
            Michael Ward
    d. Christopher B. Wyatt IV (1912–2000)
        Monica Wyatt m. Philip Burnham.

B. ST. GEORGE LEWIS LIVINGSTON CROGHAN.
B. April 23, 1822, Locust Grove.
D. November 14, 1861 [Killed in Civil War]
M. Cornelia Louisiana Livingston.

1. **Cornelia L Croghan**, born May 9, 1847, Married 1869
Horatio Horner, died 1878.
    a. Mary Sophia (1870–1935). Mary Sophia was an 8 year
    old orphan, left with Great Grandmother Serena Livingston
    Croghan who died 1885 leaving 15 year old Sophia.

2. **Lucy Serena Croghan**, born May 23, 1850, married December 11, 1866, Oakland, California, Spencer Cochrane Browne.
    a. Cornelia Ridgely Browne 1868–1882
    b. John Ross Browne 1870–1871
    c. Lucy Browne, born 1873, married Sidney Van Wyck.
    d. Florence Browne. 1875
        Spencer Browne 1885–1945, married Lena Fergurson.
        Ross Browne, Nancy Browne, Scott Browne
3. **George Croghan** (1852–1911), Inherited Locust Grove, sold in 1878. Never married
4. **Elizabeth Augusta (Lily) Croghan** (1854–1923–?) married Duncan Kennedy.
    a. Duncan Kennedy

C. SERENA CROGHAN RODGERS

B. October 17, 1834.
D. August 28, 1926.
M. Augustus Frederick Rodgers,
    1. **Daughter,** Cornelia Rodgers, 1859–1941, Married, 1881 Norval Lane Nokes
        Grand Daughter, Virginia Rodgers Nokes married John Burke Murphy.
            Great Grand Daughter, Virginia Rodgers Murphy 1905–2001,
            Married 1931 George Sterrett Wheaton
            [Presented GRC sword to Locust Grove]
                George S. Wheaton III 1933
                John Rodgers Wheaton 1938
2. **Montgomery Denison** , 1861–1901
3. **Marion St. George,** 1862–1959   D. unmarried.
4. **Nannie Louise,** 1864 –1911, D. unmarried.
5. **Augustus Frederick** 1865–1955   D. unmarried.
6. **Henry Croghan,** Married Eugenie Emanualita De Santamarina, no issue
7. **Grace Maria,** 1868–1955, D. Unmarried
8. **Robert Rodgers** married Elizabeth Renier, Robert was killed in Hong Kong during WWII fighting with the British.
    a. Robert
    b. Edward Harold m. Geraldine Cole
        Claudia Rodgers
        Christopher Rodgers
        John Rodgers
    c. Frederick married Mary
        Susan Rodgers born 1947
        John Rodgers
        Cecilia Rodgers.
        Victoria Rodgers.
        Christopher Rodgers
        Mary Ann Rodgers

IV Generation

Children of William Croghan, Jr. and Mary O'Hara

1. William Croghan III, B. March 1824. D. April 25, 1828, Pittsburgh.
2. Mary O'Hara Croghan, B. October 15, 1825—D. July 18, 1826,
Locust Grove.
3. Mary Elizabeth Croghan, B. April 13, 1827, Locust Grove
    M. January 1842, Capt. Edward Schenley,
    D. December 31, 1903, London, England.
    a. Elizabeth Pole Schenley (Lily) born June 30, 1843—(a great
    beauty)
        M. March 5, 1865, Honorable Ralph Harbord, died, 1914–15.
        Edward Harbord—In WWI captured and shut up in
        fortress when mother died.
            Had many children—He was still living at the
            beginning of WWII.
        Ida Harbord—Married Atherton Brown
        Florence Harbord Finch
        Agnes—Never Married
        Emily Harbord Des Barres
    b. Jane Inglis Schenley born 1844
        M. Reverend H. W. Crofton, Irish Parson
        Frances Hermione Crofton Cunningham
        Mary Crofton Farquharson
        Dica Crofton Forster
        Violet Crofton
        Geoffrey Crofton—(grandfather in 1940's)
    d. William Schenley born 1845–46. "Willie" died January 3, 1862
    (16 years old)
    e. Henrietta Agnes Schenley born May 1847, married Charles Ridley.
        Alberta Ridley McLean
    f. Edward Clarence Schenley Died in Boer War 1899–1902
    —South Africa
    h. Richmond Emmeline Mary Schenley married Captain Charles
    J. Randolph
        Young son—died
    i. Mellicina Isabel [Alice] Schenley married Colonel Frederic Gore
    (No Children)
    j. Lady Octavia Hermione Courtney Schenley
        Born 1858 [Mary Schenley would have been 31]
        1905, Hermione visited Pittsburgh with Nephew,
        Edward Harbord
        1906, she married Edward Downes Law, Commander of
        the British Navy, Baron of Ellenborough. Hermione
        became Baroness of Ellenborough.
    k. George Alfred Courtenay Schenley married Grace Atkinson
    Hughes
        a. Una [Kitty]—adopted one son. Una was in possession
        of Croghan Family Tea Service which was bought and
        given to Historic Locust Grove, Inc., by Mrs. Gheens.

IV Generation

Children of Ann Croghan and General Thomas Jesup

1. **Lucy Ann Jesup**
   B. Apr 17, 1823, Locust Grove, Louisville, Kentucky.
   M. February 28, 1854, Lorenzo Sitgreaves
   D. 1912, Boston, Mass.
   > Mary Sitgreaves
2. **Eliza Hancock Jesup**
   B. September 11, 1824, Fotheringay, Montgomery County Virginia.
   D. July 13, 1825 (11mos. 2 days) Washington City.

3. **Mary Serena Eliza Jesup**
   B. December 7, 1825 Washington City.
   M. January 14, 1846 James Blair, son of Blair House Fame.
   D. June 6, 1914 [President of Colonial Dames, Washington D.C. Chapter.]
   > Violet Blair married Albert Janin 1874.
   > Lucy James Blair married George Wheeler.
   > Jesup Blair
4. **Jane Findlay Jesup**
   B. November 27, 1827 Washington City
   M. Feb 1852 Augustus Nicholson
   D. 1922
   > Jesup Nicholson
5. **Elizabeth Croghan Jesup**
   B. Feb 17, 1829.
   D. June 18, 1830
6. **William Croghan Jesup**
   B. June 27, 1833 Washington City [13 Yrs. old when Mother died—1846.]
   D. November 14, 1860, Elkton, Ky. (27 yrs.4 Mos.18 days)
7. **Charles Edward Jesup**
   B. March 14, 1835, Washington City. [11 Yrs. old when Mother died—1846.]
   D. April 22, 1861 Spring Grove, near Elkton, Ky. (25 yrs. 1 Mo.)
8. **Julia Clark Jesup**
   B. July 10, 1840 [6 Yrs. old when Mother died.]
   D. 1922, Italy Washington City [Manager of Mammoth Cave, early 1880–1885.]

NOTES

Abbreviations for Some Works Frequently Cited

TFHS: The Filson Club Historical Society.

Bodley Diary: *My Dear Little Daughters Ellen and Edith,* Louisville, KY, 1903.

Temple Bodley, GRC: Temple Bodley, *George Rogers Clark—His Life And Public Service* (Cambridge: Riverside Press, 1926.)

Draper MSS: Draper Manuscript.

Dye: Eva Dye Collection, Oregon Historical Society, Portland, OR.

English: William Hayden English, *Conquest Of The Northwest,* 2 Volumes, Indianapolis, 1896.

James: James Alton James, *George Rogers Clark,* Chicago, 1928.

JC Diary: Jonathan Clark's Diary, TFHS, 1770-1811.

LC: Library of Congress.

MHS: Missouri Historical Society, St. Louis, MO.

PH: William Clark Kennerly As Told To Elizabeth Russell [1840–1870, ?], *Persimmon Hill* (Norman: University of Oklahoma Press, 1948).

PART I
The Clark and Croghan Families of Locust Grove

Chapter One
Clarks and Croghans

1. *Sunday Herald of NY City,* Mar 9, 1902.

2. William Clark Kennerly, *Persimmon Hill* (Norman: University of Oklahoma Press, 1948), p. 7.

3. Ibid., p.6

4. John H. Gwathmey, *Twelve Virginia Counties* (Richmond, Va.: The Dietz Press, 1937), p. 188.

5. Temple Bodley, *George Rogers Clark—His Life And Public Services* (Cambridge: Riverside Press, 1926), p. 4.

6. Agnes Rotherby, *Homes Virginians Have Loved* (Cambridge: Riverside Press, 1926), p. 4.

7. Jonathan Clark Diary, 1770–1811, The Filson Club Historical Society, (TFHS) Louisville, KY.

8. William Clark Kennerly, *Persimmon Hill* (Norman: University of Oklahoma Press, 1948), p. 6.

9. Draper MSS, {10J13-20}. Library of Congress, August 26, 1867

10. Samuel G. Humphries to Eva Emery Dye, March 25, 1902, Oregon Historical Society.

11. Draper MSS, (10J13-20}, August 26, 1867, LC.

12. Col. Charles Thruston to Lyman Draper, LC, [24J10].

13. Draper MSS., {10J115}.

14. J. Wyatt Jones to Eva Emery Dye, August 5–28, 1901, Oregon Historical Society.

15. Ibid.

16. Ibid.

17. The National Society of the Colonial Dames of America, Dumbarton House, Washington D.C. Letter to author, 2001.

18. J. Wyatt Jones to Eva Dye, August 28, 1901, Oregon Historical Society.

19. Ibid.

20. Ibid.

21. Charles Anderson to Lyman Draper, March 6, 1884, (10J260), LC.

22. Jonathan Clark Diary, TFHS, 1783.

Chapter Two
On the River to Kentucky

1. Jonathan Clark Diary, TFHS, Nov. 10, 1894.

2. William Hayden English, *Conquest of the Northwest,* Vol. I (Indianapolis: Bowden Co., 1896), p. 66

3. Ludie Kincaid (John Croghan Diary), The Filson Club History Quarterly, Vol. III, October 1928.

4. Shirley Streshinsky, *Audubon* (Athens: The University of Georgia Press), pp. 54–55.

5. Ernest M. Ellison, *Mulberry Hill Clark Trail Heritage Foundation* (Metro Parks, Jefferson County, KY, 2001).

6. Col. George Hancock to Lyman Draper, Draper MSS (25S213), p. 273.

7. Written By George Clark [son of Jonathan], Draper MSS (10J344-347).

Chapter Three
Mulberry Hill and Frontier Life

1. Ernest M. Ellison, *Mulberry Hill Clark Trail Heritage Foundation* (Metro Parks, Jefferson County, KY, 2001).

2. The Filson Club Quarterly, Vol. 10, p, 206.

3. TFHS, Bodley, *1903.* p. 15.

4. Keith N. Morgan. "Josiah Reconsidered: A Green Co. School Of Inlay Cabinet Making," *Antiques in Kentucky* (April, 1974).

5. TFHS, Bodley, 1903. p. 28.

6. Ibid., p. 29. William Clark inherited Mulberry Hill with all furnishings in 1802, Jefferson Co. KY, Deed Book 5, p. 481.
Isaac's sister, Ann Clark Pearce inherited "Trough Springs" from (parents) Jonathan and Sarah Hite Clark. Jefferson Co. Ky., 1811, Deed Book B 6 - p. 481.

7. Reuben T. Durrett, *The Centenary Of Louisville* (Louisville: J.P. Morton & Co., 1893), p. 5.

8. John Bakeless, *Lewis and Clark, Partners in Discovery* (New York: Morrow and Co., 1947), p. 54.

9. TFHS, Bodley, p. 45.

10. Charles Anderson, *The Story Of Soldiers Retreat,* TFHS.

11. TFHS, Bodley, 1903, p. 16.

12. William Clark Kennerly, *Persimmon Hill* (Norman: University of Oklahoma Press, 1948), p. 101.

13. Reuben T. Durrett, *The Centenary Of Louisville* (Louisville: J.P. Morton & Co., 1893), p. 48.

14. Ibid.

15. Draper MSS (10J8.)

16. TFHS, JC. Diary, Oct. 19, 1783.

17. *The Kentucky Gazette*, Vol. 3, No. 33, April 12, 1790, p. 2.

Chapter Four
The Courtly Major of Locust Grove

1. British Isles, Family Search R International Genealogical Office of Ireland: Dublin, TM V4.01, Family History Library, The Church Of Latter Day Saints, Salt Lake City, Utah.

2. 1606–1861, British Film Area-01571396, Family History Library, The Church of Latter Day Saints, Salt Lake Utah.

Some sources have incorrectly reported Nicholas Croghan of Ireland as Major Croghan's father. The parentage is based on a letter written by a Nicholas Croghan to son, "William" in America. This Nicholas and son William could have been uncle and cousin of Major Croghan. Nicholas was buyer in Europe for his relative, Indian Trader Colonel George Croghan (1772–1782). He was not the father of Major William Croghan, who married Lucy Clark. The mistaken identity refers to a will in Washington County, Pennsylvania.

3. John Croghan [London] to Mrs. Ann H. Jesup, August 18th, 1832, LC.

4. Albert T. Volwiler, *George Croghan and the Westward Movement* (Cleveland: Arthur H. Clark Co., 1926), p. 335.

5. John O'Fallon to Lyman Draper, January 18, 1847, Missouri Historical Society.

6. William Clark Kennerly, *Persimmon Hill* (Norman: University of Oklahoma Press, 1948), p. 14.

7. Joan V. Kumpitsch, (The Allaire Group, 2001). William L. Stone, *The Johnson Family* (Albany, N.Y.: Joel Munsell's Sons) p. 3.

8. William Stone, *Orderly Book of Sir John Johnson* (Albany, N.Y.: Joel Munsell, 1770).

9. Albert T. Volwiler, *George Croghan and the Westward Movement* (Cleveland: Arthur H. Clark Co., 1926), p. 290.

10. Ibid., p. 144.

11. Stefan Lorant, *Pittsburgh* (New York: Doubleday & Co, Inc., 1964), p. 34.

12. Nicholas B. Wainwright, *George Croghan Wilderness Diplomat* (Chapel Hill: University of North Carolina Press, 1959), pp. 264–265

13. William Croghan to Speaker of General Assembly of Virginia, 1784, LC.

14. Albert T. Volwiler, *George Croghan and the Westward Movement* (Cleveland: Arthur H. Clark Co., 1926), p. 144.

15. Rabbi David Philipson, *Letters of Rebecca Gratz* (Philadelphia: The Jewish Publication Society of America, 1929).

16. Journal of William Croghan, 1779–1780, Draper MSS, (3N1-134).

17. William Croghan to Bernard Gratz, April 22, 1779 (Lancaster, PA), Historical Society of Pennsylvania.

18. Ibid., February 11, 1780.

19. John O'Fallon to Lyman Draper, January 18, 1847, Draper MSS, (34J9) Missouri Historical Society.

20. William Croghan (Charlestown, SC), June 10, 1780 to (Uncle) George Croghan, Lancaster, PA, Historical Society of Pennsylvania.

21. H. Graff and John A. Krout, *The Adventure of the American People* (Skokie, Ill.: Rand McNally & Co., 1971), p. 95.

22. James A. James, *The Life Of George Rogers Clark* (Urbana: University of Chicago Press, 1928), pp. 246–249.

23. Reuben T. Durrett, *The Centenary Of Louisville* (Louisville: J.P. Morton & Co., 1893), p. 96.

24. Jack Harrison, "Colonel George Rogers Clark and Co." Louisville Area Chamber of Commerce (January, 1978), p. 49.

25. Petition of William Croghan to Speaker of General Assembly or VA, 1784, Wisconsin Historical Society.

26. Albert T. Volwiler, *George Croghan and the Westward Movement* (Cleveland: Arthur H. Clark Co., 1926), p. 332.

27. William Croghan [Fort Pitt] to Colonel Davis, August 18, 1781, Virginia State Library.

28. Charles Anderson to Charles E. Rice, June 8 1894, TFHS.

29. Boynton Merrill, Jr., *Jefferson's Nephews* (Princeton: Princeton University Press, 1976), p. 121.

30. Ibid., p. 95.

31. Ibid., Thomas Jefferson to Lucy Jefferson Lewis, Washington, April 19, 1808, pp. 148–149.

Chapter Five
Locust Grove

1. Chenoweth Massacre, *History Of The Massacre Trail* (Historic Middletown, Inc., Troop 321, B.S.A.)

2. George Hancock to Lyman Draper, November 20, 1868, LC.

3. William Croghan's surveyor's office at house near Falls of Ohio, *The Kentucky Gazette*, April 26, 1791.

4. George Hancock to Lyman Draper, November 20, 1868, LC.

5. The Register of Kentucky State Historical Society, Frankfort, KY, Vol. 22, No. 66, 1924.

6. John O' Fallon (St. Louis) to Lyman Draper, January 18, 1847: August 6, 1847, Missouri Historical Society.

7. Draper MSS (4CC172), John Clark to James O'Fallon, May 28, 1792.

8. Draper MSS (25S213), John O'Fallon (St. Louis) to Lyman Draper, January 18, 1847.

9. Alfred Tischendorf, Editor, *Diary and Journal of Richard Clough Anderson, Jr.* (Durham: Duke University Press, 1964).

10. Lawrence L. Barr, *A New Look at the History of Soldiers Retreat* (July 25, 1979).

11. Charles Anderson, *The Story of Soldiers Retreat*, TFHS, Louisville, KY.

12. Katherine Jennings, *Louisville's First Families* (Louisville: Standard Press Co.) p. 118.

13. John O'Fallon to Lyman Draper, August 6, 1847, Draper MSS (34J16-19), Missouri Historical Society.

14. Ibid.

15. The Filson Club Historical Quarterly, Vol. 6, July 24, 1799.

16. William Clark to John O'Fallon, November 22, 1808, Missouri Historical Society.

17. John O'Fallon (Louisville, KY) to Gen. George R[ogers] Clark, Nov. 18, 1808, Draper MSS (55J64), Missouri Historical Society.

18. Ibid., (24J8-12) February 1847.

19. Frances Clark Fitzhugh to William Clark [No Date]. Missouri Historical Society.

20. Rector James Craile, *Sketches Of Christ Church Cathedral* (Louisville: J.P. Morton & Co., 1862).

21. John H. Gwathmey, *Twelve Virginia Counties* (Richmond, Va.: The Dietz Press, 1937), p. 235.

22. Board of Commission Allotment Of Land In Clark's Grant, Dec., 1808.

23. The Filson History Quarterly, 1969, [Vol.43], p. 50.

24. George Rogers Clark to General. W. [William] Clark, St. Louis, October 27, 1811, Missouri Historical Society.

25. *The Western Pennsylvania Historical Magazine*, Vol. 51, #3, July 1968.

26. The Filson History Quarterly, 1969, [Vol. 43], p. 50.

Chapter Six
Explorers Leave the Falls

1. Nicholas Biddle, Edited, *Lewis and Clark Letters*, #412, December 4, 1783, pp. 654–655.

2. Ibid., #413, February 8, 1784, pp. 655–656.

3. Robert B. Betts, *In Search Of York* (Boulder: Colorado Associated University Press, 1985), p. 101.

4. William Clark Kennerly, *Persimmon Hill* (Norman: University of Oklahoma Press, 1948), p. 11

5. Stephen Ambrose, *Undaunted Courage* (New York: Simon & Schuster, 1996), p. 57.

6. Ibid.

7. Donald Jackson, *Thomas Jefferson & The Stony Mountains* (Champaign, Ill.: University of Illinois Press, 1981), p. 139.

8. David Lavender, *The Way The Western Sea* (New York: Harper & Row, 1817), p. 378.

9. Eva Emery Dye, *The Conquest* (New York: Wilson-Erickson, Inc., 1936), p. 219.

10. William Clark (St. Charles, MO) to Maj. Wm. Croghan, May 21, 1804, Nicholas Biddle, Editor of Lewis and Clark letters, p. 195. (Eva Emery Dye traveled East, interviewed families and copied family letters.)

11. *Early Louisville, KY Newspaper Abstracts*, Clark's brother-in-law Dennis Fitzhugh operated a store in the vicinity of 5th and Main Sts. William Clark shopped there on November 5, 1806, the day he, Lewis, York, and others arrived at the Falls of Ohio.

12. William Clark Kennerly, *Persimmon Hill* (Norman: University of Oklahoma Press, 1948), pp. 19–20.

13. Ibid.

14. Richard Dillon, *Meriwether Lewis* (New York: Coward-McCann, Inc., 1965), p. 265.

15. From Eva Emery Dye Collection, Oregon Historical Society.

16. Oregon Historical Society, Letter sent to Melzie Wilson, 2000.

17. Robert B. Betts, *In Search Of York* (Bolder: Colorado Associated University Press, 1985), pp. 418–419.

18. William Hancock Clark to Eva Emery Dye, (1) May 21, l901; (2) May 10, 1905, Dye Collection, Oregon Historical Society.

19. C. Harper Anderson to Eva Emery Dye, 1901, Oregon Historical Society.

Meriwether Lewis's nephew C. Harper Anderson wrote interesting letters about his Uncle Meriwether. In 1901, the family had the spy glass used on the Expedition journey and "also, his

Masonic Apron with which my Mother protected herself at home from a rabble of drunken Union Soldiers, during Sheridan's Raid through this Section in March 1865."

20. John Bakeless, *Lewis and Clark, Partners in Discovery* (New York: Morrow and Co., 1947), p. 376.

21. Richard Dillon, *Meriwether Lewis* (New York: Coward-McCann, Inc., 1965), p. 265.

22. William Clark to Major W. Croghan, Dec. 14, 1806, Historical Society of Pennsylvania.

23. John Bakeless, *Lewis and Clark, Partners in Discovery* (New York: Morrow and Co., 1947), p. 383.

24. Mrs. Kennerly Taylor to Eva Emery Dye (No Date), Dye Collection, Oregon Historical Collection.

25. Eva Emery Dye Collection, Oregon Historical Society, Original missing from Dye Envelope.

26. William Clark Kennerly, *Persimmon Hill* (Norman: University of Oklahoma Press, 1948), p. 25.

27. Edmund Clark (Henderson, KY) to Jonathan Clark, May 1808, TFHS, Edmund Clark File.

28. George Rogers Clark Sullivan to John O'Fallon, June 2, 1808, Missouri Historical Society.

29. John Bakeless, *Lewis and Clark, Partners in Discovery* (New York: Morrow and Co., 1947), pp. 390–391.

30. Ibid.

31. Ibid.

32. Pete H. Clark (St. Louis) to Mrs. Eva Dye, (Oregon City), December 6, 1900.

33. William Clark Kennerly, *Persimmon Hill* (Norman: University of Oklahoma Press, 1948), p. 31.

34. Robert B. Betts, *In Search Of York* (Boulder: Colorado Associated University Press, 1985), pp. 112–114.

35. George Hancock to Wm. Croghan, December 15, 1828, Darlington Memorial Library, Pittsburgh.

36. Dumas Malone, *Jefferson the President* (New York: Little, Brown & Co., 1974), pp. 207–208.

Chapter Seven
General George Rogers Clark

1. George Gwathmey received request from General George Rogers Clark (Uncle), July 1808, Draper MSS (55J63).

2. Dumas Malone, *Jefferson the President* (New York: Little, Brown & Co., 1974), p. 207.

3. Edmund Clark (Henderson, KY) to Jonathan Clark, May 1808., TFHS.

4. Temple Bodley, *George Rogers Clark—His Life And Public Services* (Cambridge: Riverside Press, 1926), p. 335.

5. *American State Papers,—Public Lands*, p. 1,274, (Draper MSS, 54J49) LC.

6. Ibid.

7. Temple Bodley, *George Rogers Clark—His Life And Public Services* (Cambridge: Riverside Press, 1926), p. 331.

General G.R. Clark's vouchers were a remarkably correct account of his last five years. At that time, false charges were being brought to Governor Harrison and circulated at Richmond which stated that he was 'a sot' and incapable of attending to business. In 1913, one hundred thirty-one years after he received the auditor's receipt for his vouchers, seventy large packages of those originals were found in a room of the auditor's building in Richmond. They had not been destroyed by Arnold's British soldiers.

8. John Bakeless, *Background To Glory* (Philadelphia: Lippincott Co., 1957), p. 348.

9. William Clark (St. Louis) to Jonathan Clark, November 22–24, 1808, TFHS, Clark Collections.

10. Jerome O. Steffen, *William Clark* (Norman: University of Oklahoma Press, 1942), p. 29.

11. TFHS Quarterly, Vol. 2, October 1927. No. 1. Charles K. Needham, *A Review Of The Efforts To Develop Water Power At the Falls Of The Ohio*, p. 3.

12. James, New York: Ams Press, Inc. 1928, p. 457.

13. Jefferson County, KY, 1811, Deed Book B 6 – p. 48.

14. Edmund Clark [Henderson, KY] to Jonathan Clark, May 1808, TFHS.

15. Dumas Malone, *Jefferson the President* (New York: Little, Brown & Co., 1974), pp. 438–39.

16. Ben Cassady, *History of Louisville* (Louisville: Hull and Brother, 1852), pp. 34–35.

17. *The Boston Post*, TFHS.

18. Draper MSS (24J10-13).

19. Written by Ann Croghan Jesup's Granddaughter, [Not Dated] Violet Blair Janin, Huntington Library, California.

20. Isabel McLennan McMeekin, *The Gateway City* (New York: Messner Inc., 1946), p. 71.

21. Draper MSS (10J307), LC.

22. Temple Bodley, *George Rogers Clark—His Life And Public Services* (Cambridge: Riverside Press, 1926), p. 365.

23. R. W. Ferguson to Lyman Draper, August 27, 1869, Draper MSS (35J8).

24. Mrs. Ann C. T. Farrar, August 15, l869, Draper MSS (10J08), LC.

Chapter Eight
1811 Earthquake and River Trade

1. Ben Cassady, *History of Louisville* (Louisville: Hull and Brother, 1852), p. 120.

2. John O' Fallon to Lyman Draper, (234J8-12). Wisconsin Historical Society.

3. Temple Bodley, *George Rogers Clark—His Life And Public Services* (Cambridge: Riverside Press, 1926), p. 366.

4. Ben Cassady, *History of Louisville* (Louisville: Hull and Brother, 1852), pp. 122–125.

5. Ibid.

6. Charles Anderson, *The Story of Soldier's Retreat*, Part I, p. 1, TFHS.

7. Jefferson County Minutes, Book 2, p. 8l.

8. From Mrs. McCauley Smith's Locust Grove papers given to Education Committee.

9. Samuel W. Thomas, *The Restoration of Locust Grove* (Louisville: 1983)

10. Charles Henry Ambler, *A History of Transportation in the Ohio Valley, Clark Co.* (Glendale, Calif.: 1932), p. 39.

11. Charles Anderson, *The Story of Soldier's Retreat*, TFHS.

12. William Croghan [Danville] to Richard Clough Anderson, [near Louisville], November 20, 1788. LC.

13. H. M. McMurtrie, *Sketches Of Louisville* (Louisville: S. Penn, Jun., Main Street, 1819) p. 119.

14. Melville O. Briney, *Fond Recollections* (Louisville: *Louisville Times*, 1969), p. 38.

15. Jerome O. Steffen, *William Clark* (Norman: University of Oklahoma Press, 1977), p. 23.

16. Jon Kukla, *A Wilderness so Immense* (New York: Alfred A. Knopf, 2003), p. 166.

17. Ibid., Governor Isaac Shelby of Kentucky, January 13, 1794, p. 178.

18. Mrs. Eleanor Eltings Temple to Lyman Draper, August 16–17, 1867. Draper MSS (10J5512).

19. Donald Jackson, *Thomas Jefferson and The Stony Mountains* (Urbana: University of Illinois Press, 1981), p. 153.

20. George Hancock to Lyman Draper, November 20, 1868. Wisconsin Historical Society.

21. Dedlake Gresham & Dedlakes account and receipt for General Clark's Carriage, $310, August 7, 1813. Draper MSS (54J67).

22. Ibid., (54J54).

23. Alice Ford, Editor of John James Audubon Writings, *Audubon, by Himself* (New York: The Natural History Press, 1969), pp. 16–17.

24. Francis Vigo to Gen. George Rogers Clark, July 15, 1811. Draper MSS (55J77)

25. Col. George and John Croghan as told to Lyman Draper, November 8, 1844. Draper MSS (10J204)

Chapter Nine
War of 1812

1. Eva Emery Dye, *The Conquest* (New York: Wilson-Erickson, Inc., 1936), p. 369.

2. Ibid., 370–371.

3. Ibid.

4. Charles Richard Williams, Ph.D., *Major George Croghan* (Columbus, Ohio: Press of Fred J. Heer, 1903), pp. 387–391.

5. To Ms. Emilie Smith from Lelia Newhall, (1960s), (Granddaughter of George and Serena Croghan, California).

6. John O'Fallon to Draper, February 1847, Wisconsin Historical Society.

7. Alfred Tischendorf, Editor, *Diary and Journal of Richard Clough Anderson*, Jr. (Durham: Duke University Press, 1964), p. 39.

8. Nicholas Biddle, Edited *Lewis And Clark Letters,* p. 600.

9. *The Louisville Evening Post*, Friday, February 13, 1818.

Chapter Ten
A Day in the Life of Lucy

1. Ben Cassady, *History of Louisville* (Louisville: Hull and Brother, 1852), p. 140.

2. Samuel W. Thomas, *The Restoration of Locust Grove* (Louisville: 1983), p. 149.

3. J. Wyatt Jones, August 5–28, 1901, Dye Collection, Oregon Historical Society.

George Washington's Mount Vernon comes into mind when writing about the lawns at Locust Grove. In *Mount Vernon Ladies Association*, a good description of gardens and lawns is found. The landscaped area of garden and lawns about the mansion is separated from the surrounding fields on three sides by sunken walls or 'ha-haw'. The bowling green entrance, a flanking 'ha-haw' wall mark the boundary on the west between the formal and the informal areas. Washington wrote, "I do not hesitate to confess, that reclaiming and laying the grounds down handsomely to grass, and in woods thinned, or in clumps, about the Mansion house is among my first objects and wishes". William Clark Kennerly, *[William Clark's nephew]* writes about a similar approach to his home in St. Louis: "The approach to this door, through a wide portico, was by a well-graveled driveway circling from front and back to the distant gate that swung between two high pillars of rough stone." p. 70.

4. Samuel W. Thomas, *The Restoration of Locust Grove* (Louisville: 1983), p. 2.

5. Jefferson Co. Tax Records, 1816

6. "Alfred Beckley's Recollections of KY," The Register of KY Historical Society Vol. 60, No. 4 (1962), pp. 304–313.

7. Jane C. Nylander, *Early American Life* (October 1980), p. 46.

8. Jerome O. Steffen, *William Clark* (Norman: University of Oklahoma Press, 1977), p. 17.

9. Frederick Way, Jr., *The Allegheny* (New York: Farrar & Rinehart, 1942), p. 92.

10. John Croghan [London] to Ann Jesup, August 18, 1832. LC.

11. Edward O. Welles, "George Washington Slept Here," *Historic Preservation* (May/June 1982), p. 26.

12. William Seale, *The President's House,* White House Historical Association (Vol. I), p. 129.

13. While conducting a Locust Grove tour [1977], Ms. Lyons Brown explained the purpose of a narrow trench in the center of the walk which lay alongside the outside kitchen wall.

14. The National Society of the Colonial Dames of America. Washington D.C.

15. Marguerite Ikes, *The Standard Book Of Quilt Making and Collecting* (New York: Dover Publishing, 1959).

16. John Croghan to Genl. Ths. S. Jesup, April 10, 1825, LC.

17. Amy La Follette Jenson, *The White House and Its Thirty-Four Families* (New York: McGraw Hill, 1958), p. 23.

18. Manuscript & Archives Division, The New York Public Library.

19. Alfred Tischendorf, Editor, *Diary and Journal of Richard Clough Anderson,* Jr. (Durham: Duke University Press, 1964), p. 88.

20. William Walker, *Southern Harmony* (Lexington, Ky.: University of Kentucky Press, 1927).

21. Saul K. Padover, *A Jefferson Profile* (New York: John Day Co.), p. 298.

22. David McCullough, *John Adams* (New York: Simon and Schuster, 2001), p. 104.

23. Charles Anderson, *The Story Of Soldiers Retreat,* (1814–1895), p. 4.

24. William Clark Kennerly, *Persimmon Hill* (Norman: University of Oklahoma Press, 1948), pp. 52–53.

25. Ibid.

26. Ibid., pp. 53–54.

27. John Bakeless, *Lewis and Clark, Partners in Discovery* (New York: Morrow and Co., 1947), p. 447.

28. Ibid.

29. Jerome O. Steffen, *William Clark* (Norman: University Of Oklahoma Press, 1942), p. 151.

30. Nicholas Biddle, Edited *Lewis and Clark Letters*, William Clark L.G to Nicholas Biddle, Letter #404, p. 638.

31. Judge Dennis Fitzhugh to John O'Fallon, October 16, 1821. Missouri Historical Society.

32. John Bakeless, *Lewis and Clark, Partners in Discovery* (New York: Morrow and Co., 1947), p. 447.

33. William Clark [Locust Grove] to Nicholas Biddle, Letter No. 404, p. 638.00

34. William Clark Kennerly, *Persimmon Hill* (Norman: University of Oklahoma Press, 1948), p. 56.

35. Ibid., p. 65.

Chapter 11
Social Life—Matchmaking—Marriage

1. Jefferson County Minute Book 2, p. 81.

Major Croghan served in State Constitutional Convention, establishing Kentucky Statehood, 1792. He was one of original trustees of City of Louisville and Jefferson Seminary [University of Louisville].

2. David Rozel Poignand, 1869, Draper MSS (35S78).

3. Ben Cassady, *History of Louisville* (Louisville: Hull and Brother, 1852), pp. 37–38.

4. Isabel Mclennan McMeekin, *Louisville, The Gateway City* (New York: Messner, Inc., 1946), p. 188.

5. Garraty and Cornet, Editors, *The World Book Encyclopedia*, Vol. 6 (Chicago: Field Enterprises, Inc., 1959), p. 2826.

6. Ben Cassady, *History of Louisville* (Louisville: Hull and Brother, 1852), pp. 50–51.

7. Isabel Mclennan McMeekin, *Louisville, The Gateway City* (New York: Messner, Inc., 1946), p. 188

8. J. Wyatt Jones to Eva Emery Dye, August 5–28, 1901, Oregon Historical Society.

9. William Clark to Major Croghan, May 18, 1817, Historical Society of Pennsylvania.

10. Alfred Tischendorf, Editor, *Diary and Journal of Richard Clough Anderson*, Jr. (Durham: Duke University Press, 1964), p. 56.

11. *The Courier Journal*, Louisville, Kentucky, June 26, 1819.

12. *The Courier Journal*, Louisville KY., June 26, 1819.

13. Stoddard Johnston to Eva Emery Dye, September 10, 1901, Oregon Historical Society.

14. Ibid.

15. Janin Collection, Philip Lee to Violet, [No Date] HL.

16. William Clark Kennerly, *Persimmon Hill* (Norman: University of Oklahoma Press, 1948), p. 255.

17. Jefferson County Deed Book 6, p. 544.

18. See Part IV, Chapter One, Note #24.

19. Hancock to Jesup, August 28, 1820, LC.

20. George Hancock [Fotheringay] to Jesup, August 6,1821, LC.

21. John Croghan to Jesup, Washington City, March 11, 1822, LC.

22. John B. Larner (Editor), *Columbia Historical Society, Washington, D.C.*, 1928, Vol. 2, p. 155

23. William Croghan Jr., to Major William Croghan, May 18, 1822, Historical Society of Western PA.

24. John O'Fallon to Dennis Fitzhugh, May 10, 1822. MHS.

25. Wallpaper on the ball room wall at Locust Grove is a reproduction of the original. When removing a wall during restoration [1960–1964], scraps of the original paper were found. Research led to the Reveillon Studio in Paris, France, where the paper was originally produced, and they were able to reproduce the exact paper for Locust Grove.

26. Amy La Folette Jensen, *The White House and Thirty Four Families* (New York: McGraw Hill, 1965), p. 23.

27. Mary Croghan to Lucy A. Jesup, August 30,1833, Croghan Letter, LC.

28. Catharine S. Zimmerman, *The Bride's Book*.

29. *Colonial Homes: Artists In Aprons,* March-April, 1979, p. 128: *Womanly Arts,* July-August, 1985.

Chapter Twelve
Eventful Year for Lucy

1. Ben Cassady, *History of Louisville* (Louisville: Hull and Brother, 1852), p. 140.

2. Edmund Croghan's last letter, July 17, 1822, LC.

3. Major Croghan died September 21, 1822, *Louisville Public Advertiser*, Vol. 4, No. 13, p. (3), col. (5).

4. Alfred Tischendorf, Editor, *Diary and Journal of Richard Clough Anderson,* Jr. (Durham: Duke University Press, 1964), p. 6.

5. George Hancock [Richmond] to Brig. Genl. Thomas S. Jesup, [Washington City,] December 25, 1822, LC.

6. *Pittsburgh Mercury*, February 4, 1823, Vol. X1, No. 553, p. 2, Col. 13.

7. John Croghan [Louisville] to General. T.S. Jesup [Washington], April 10, 1825, LC.

8. Ibid., May 20, 1825.

9. Ben Cassady, *History of Louisville* (Louisville: Hull and Brother, 1852), p. 165.

10. John Croghan [Louisville] to General. T.S. Jesup [Washington], May 20, 1825, LC.

11. O'Fallon Family Records, MHS.

12. Alfred Tischendorf, Editor, *Diary and Journal of Richard Clough Anderson,* Jr. (Durham: Duke University Press, 1964).

13. Major Gist Blair, *Lafayette Square*, 1926, Vol. 28, p. 151, Columbia Historical Society.

14. Alfred Tischendorf, Editor, *Diary and Journal of Richard Clough Anderson,* Jr. (Durham: Duke University Press, 1964).

15. Gov. William Clark [Washington City] to My Dear Son, February 18, 1829, *Persimmon Hill*, p. 57.

16. Amy La Folette Jensen, *The White House and Thirty Four Families* (New York: McGraw-Hill, 1965), p. 55.

17. John Croghan to Genl. Ths. S. Jesup, March 15, 1825, LC.

18. John R. Livingston [Red Hook, NY] to William Croghan, [Jr.] July 17, 1827, Darlington Memorial Library.

19. William Croghan, Jr. to Thos. S. Jesup, August 28, 1827, LC.

20. Alfred Tischendorf, Editor, *Diary and Journal of Richard Clough Anderson,* Jr. (Durham: Duke University Press, 1964). p. 187.

21. Ibid.

22. Ibid. [Diane Gwathmey Bullitt's son also died in Columbia, February 18, 1824].

Chapter Thirteen
Last Days at Locust Grove

1. *Pittsburgh Mercury,* Vol. SVI, No. 827, p. [3], col. [3], May 6, 1828.

2. John Croghan to Mrs. Ann H. Jesup, August 18, 1832, London, LC.

3. John Croghan to My Dear Gene. [Thomas Jesup] October 17, 1832, Paris, France. LC.

4. George Hancock to Gen. Tho. S. Jesup, June 24, 1833, Locust Grove. LC.

5. Ibid., June 30, 1833.

6. William Croghan to Jesup, July 10, 1834. LC.

7. Farmers Fire Insurance, NY, August 13, 1831, $15,000 for George Hancock, Locust Grove. Jefferson Co. Ky , $7,000 for House, $4,000, for Private Library, $15,000.

8. John Croghan to Jesups, March 5, 1835. LC.

9. John Croghan to Jesup, August 20, 1837. LC.

10. William Clark Kennerly, *Persimmon Hill* (Norman: University of Oklahoma Press, 1948), pp. 90–91.

11. Ibid., 115.

12. Christopher Graham to Lyman Draper, November 15, 1882. LC.

13. Colonel Reuben T. Durrett, *The Courier Journal*, July 10, 1883.

14. John O'Fallon [St Louis] to Joseph Taylor, January 29, 1849.

15. Joseph Taylor [Baltimore] to John O'Fallon [St. Louis] March 20, 1849, Missouri Historical Society.

PART II
Born to Be a Soldier

Chapter One
The Livingstons

1. Garraty and Cornet, Editors, *The World Book Encyclopedia,* Vol. 10 (Chicago: Field Enterprises, Inc., 1959), p. 4540.

2. Reuben Hyde Walworth, *The Livingston Genealogy (Rhinebeck, N.Y.:* Historical Society Collections, 1982).

3. Abstracts of Wills on file in The Surrogates Office, City of New York, NY, Historical Society Collections, 1903, XII. [NY 1906]

4. Lewis Paul Todd and Merle Curti, *Rise Of The American Nation* (New York: Harcourt, Grace & World, Inc., 1966), p. 56.

5. Reuben Hyde Walworth, *The Livingston Genealogy (Rhinebeck, N.Y.:* York Historical Library, 1982), p. 40.

6. George Dangerfield, *Chancellor Robert R. Livingston* (New York: Harcourt, Brace & Co., 1969), p. 97.

7. Ibid., p. 112.

8. Ibid., p. 112.

9. Ibid., p. 113.

10. Ibid., p. 128.

Chapter Two
British March on Livingston Manor

1. George Dangerfield, *Chancellor Robert R. Livingston* (New York: Harcourt, Brace & Co., 1969), p. 81.

2. Edgar Mayhew Bacon, *Chronicles Of Tarrytown And Sleepy Hollow* (New York: Putnam, 1897), p. 73.

3. Clare Brandt, *An American Aristocracy The Livingstons* (New York: Doubleday), p. 122.

4. Edgar Mayhew Bacon, *Chronicles Of Tarrytown And Sleepy Hollow* (New York: Putnam, 1897), p. 79.

5. Clare Brandt, *An American Aristocracy The Livingstons* (New York: Doubleday), p. 125.

6. Chancellor Robert to John Livingston, Livingston, NY, April 11, 1778, New York Public Library.

7. George Dangerfield, *Chancellor Robert R. Livingston* (New York: Harcourt, Brace & Co., 1969), p. 112.

8. Reuben Hyde Walworth, *The Livingston Genealogy* (New York: Historical Society Collections, 1982).

9. George Bancroft, *Life of Edward Livingston* (New York: Appletons, 1964), p. 59.

10. James D. Livingston and Arthur Kelly, *A Livingston Genealogy*, 1986, compiled for The Friends of Clermont, Sojourner Truth Library State University, New York. Co-sponsored by The Order of Colonial Lords of Manors in America.

11. C. Edward Skeen, *John Armstrong, Jr.* (New York: Syracuse University Press, 1981), p. 38

12. George Croghan, III to Albert Janin, Janin Family (Blair), Collection, HL.

13. Margaret Beekman Livingston to RRL, Clermont, June 4, 1782, RRLP.

14. Edgar Mayhew Bacon, *The Hudson River*, 1910 (Washington Irving to Henry Brevoort, 1812) Knickerbocher Press, N.Y., p. 250.

15. Jane Wyatt to Melzie Wilson, 2002.

16. Fitzhugh Lee to Violet Janin, February 20, 1870, HL. New York 1810 census records. Henry Lee of "Stradford", Westmoreland County, Virginia, was living in Columbia Co., New York near the Livingstons.

17. C. Edward Skeen, *John Armstrong, Jr.* (New York: Syracuse University Press, 1918) p. 212.

18. Paul C. Nagel, *The Lees of Virginia* (New York: Oxford, 1990).

Chapter Three
Born to Be a Soldier

1. Charles Richard Williams, *Address Delivered Before the George Croghan DAR Chapter*, August 1, 1903, p. 384, *The Ohio Historical Society.*

2. Ibid.

3. MHS, G. R. Clark to William Clark, October 7, 1811, O'Fallon Papers.

4. Charles Richard Williams, *Address Delivered Before the George Croghan DAR Chapter*, August 1, 1903, p. 385, Ohio Historical Society.

5. Ibid., p. 386.

6. Ibid., p. 388.

7. Freeman Cleaves, *Old Tippecanoe* (Newton, Conn.: American Political Biography Press, 1996), p. 181.

8. Ibid.

9. Charles Richard Williams, *Address Delivered Before the George Croghan DAR Chapter*, August 1, 1903, p. 388, Ohio Historical Society.

10. Ibid., p. 391.

11. Charles Richard Williams, *Address Delivered Before the George Croghan DAR Chapter*, August 1, 1903, Ohio Historical Society.

12. Major William Croghan to William Davies, June 6, 1782, Illinois State Historical Library, Vol. 19, Springfield, 1924.

13. Charles Richard Williams, *Address Delivered Before the George Croghan DAR Chap*ter, August 1, 1903. p. 393, Ohio Historical Society.

14. Anderson C. Quisenberry, *Kentucky In The War of 1812* Kentucky Historical Society, 1969, p. 69.

Chapter Four
George and Serena

1. Walworth, Reuben Hyde, *Livingston Genealogy*, March, 1982, Rhinebeck, NY, # IV, p. 4l.

2. Dunlap, William, *The Arts of Design* (New York: Dover Pub., reprint of the original 1834 Edition, p. 84.

3. Stern, Philip Van Doren, *Robert E. Lee* (New York: McGraw-Hill Book Co. Inc., 1963), p. 45.

4. Major William Croghan to William Clark, May, 19, 1816, LC.

5. Virginia Wheaton (Pasadena, CA) to Melzie Wilson, (Louisville, KY.) March 2, 1988.

6. Freeman Cleaves, *Old Tippecanoe* (Newton, Conn.: American Political Biography Press, 1996), p. 246.

7. Anderson C. Quisenberry, *Kentucky In The War of 1812* (Kentucky Historical Society, 1969), p. 221.

8. Serena L. Croghan to Rutherford B. Hayes, March 22, 1872, Rutherford B. Hayes Papers, Hayes Library., Fremont, OH.

9. George Croghan, Official Army Reports, 1825–1845, National Archives.

10. Ibid.

11. George Croghan, (N.Y.) February 27, 1817, to Thomas Jesup, LC.

12. Alfred Tischendorf, Editor, *Diary and Journal of Richard Clough Anderson*, Jr. (Durham: Duke University Press, 1964).

13. William Clark to Major William Croghan, May 18, 1817, LC.

14. Mrs. Alice Newhall O'Meara to Mrs. Emilie Smith, 1960s.

15. Dunlap, William D., *The Arts of Design,* Vol. 2, Pt. I. (New York: Dover, 1834), p. 84.

16. Mary L. Winn to Lyman Draper, July 21, 1882.

17. Orlando Brown to John Brown, July 28, 1828, Brown Papers, Yale University Library.

18. George Bancroft, *Life of Edward Livingston* (New York: Appleton, 1954), p. 30.

Chapter Five
New Orleans

1. John Croghan to Brigadier General. Thomas. S. Jesup, March 15, 1825, LC.

2. John R. Livingston to William Croghan, Esq., July 17, 1827, Darlington Memorial Library, Pittsburgh.

3. Alfred Tischendorf, Editor, *Diary and Journal of Richard Clough Anderson*, Jr. (Durham: Duke University Press, 1964).

4. Ibid., p. 42.

5. Ibid., October14, 1818, p. 90.

6. Serena Livingston Croghan to Lucy Croghan, December 7, 1826.

7. John R. Livingston to William Croghan, July 17, 1827, Darlington Memorial Library, Pittsburgh.

8. George Croghan (NY) to Brig. Gen. Jesup, May 13, 1828, LC.

9. Garraty and Cornet, Editors, *The World Book Encyclopedia,* Vol. 8 (Chicago: Field Enterprises, Inc., 1959), p. 3292.

10. Freeman Cleaves, *Old Tippecanoe* (Newton, Conn.: American Political Biography Press, 1996), pp. 246–247.

11. B. Elbert Smith, *Francis Preston Blair* (New York: Free Press, 1980), p. 133.

12. Freeman Cleaves, *Old Tippecanoe* (Newton, Conn.: American Political Biography Press, 1996), p. 246.

General (Senator) Harrison in Philadelphia at public dinner, made toast in which he said, the victory of Fort Stephenson was not so much to a gallant defense as to the enemy's "blindness & folly ... McAfee's 'Late War' was devoid of the truth, as you well know, Sandusky could have been taken and that the enemy acted stupidly."

13. Ibid., p.180.

14. Francis Paul Prucha, Editor, *Army Life On The Western Frontier,* Norman: University of OK Press. XXI.

15. George Croghan to Thomas Jesup, November 19, 1825, LC.

16. Ibid., December 20, 1825.

17. Robert B. McAfee, *History of the Late War in the Western Country* (Lexington, Ky.: 1816).

18. James Preston Blair, *The Globe,* Volume 6, No. 10, August19, 1840.

19. Freeman Cleaves, *Old Tippecanoe* (Newton, Conn.: American Political Biography Press, 1996), p. 244.

20. Ibid.

21. James Preston Blair, *The Globe,* Vol. 6, No. 10, August 19, 1840.

22. Freeman Cleaves, *Old Tippecanoe* (Newton, Conn.: American Political Biography Press, 1996), p. 342.

23. Ibid., p. IX.

Chapter Six
Inspector General on the Frontier

1. Col. Wm. Lindsay to Gen. Jesup, August 1828, LC.

2. George Croghan to Thomas Jesup, April 13, 1830, LC.

3. George Croghan to Major General Macomb, National Archives, August 6, 1830.

4. Serena Croghan to Gen. Jesup, May 9, 1835, LC.

5. Anne Jesup to Gen. Jesup , February 17, 1839, LC.

6. John Croghan to Mrs. George Croghan, May 23, 1838, HL.

7. George Hancock to Gen. Jesup, December 15, 1838, LC.

8. TFHC Quarterly, Vol. 6, p. 301.

9. Ibid., p. 339.

10. Jefferson Co. Ky. Bk. 3, p. 215. January 12, 1830.

11. Thomas. S. Jesup to Dr. Croghan, September 13, 1840, LC.

12. *Magazine of Virginia Genealogy,* May 1995 #2, Vol. 33. Published by The Virginia Genealogical Society.

13. Jefferson Co: 15 Oct. 1844, Filed in Clerks Office of Ballard on 3 Dec. 1844 & Recorded therein on 12 Jan. 1845.

14. William Croghan to Ann Jesup, August 9, 1835, LC.

15. Ibid.

16.Virginia Wheaton to Emilie Smith. [1960s].

17. William Clark Kennerly, *Persimmon Hill* (Norman: University of Oklahoma Press, 1948), p. 28.

18. George Hancock to Dr. John Croghan, January 16, 1841, LC.

19, John Croghan to William Croghan [Nephew], January 16, 1846, LC.

20. John Croghan to Genl. Jesup, December 25, 1844, LC.

21. John Croghan to Genl. Jesup, February 2. 1945, LC.

22. George Croghan [Washington] to Dr. [John Croghan] March 22, 1845, University of Michigan.

23. National Archives, War Dept. April, 1844.

24. C-SPAN Archives, [American President Series] *Life Portrait of William Henry Harrison*, May 10, 1999, West Lafayette, IN.

25. John O'Fallon, Missouri Historical Society Collection.

26. Francis Paul Prucha, Editor, *Army Life on the Western Frontier*, Selections from the Official Reports (1826–1845) of Col. George Croghan (Norman: University of Oklahoma Press).

27. Henry & Jesup, *Edward Jesup of Green's Farms* (Cambridge: John Wilson & Son,1887, pp. 152–154.

28. Col. George Croghan [Philadelphia] to Wm. A. Gordon, Esq., Washington DC, office of Quarter Master General, February 1844, LC.

29. Garraty and Cornet, Editors, *The World Book Encyclopedia,* Vol. 12 (Chicago: Field Enterprises, Inc., 1950), p. 5984.

30. Francis Paul Prucha, Editor, *Army Life on the Western Frontier* (Norman: University of Oklahoma Press), Reports between 1826–1845.

Chapter Seven
Mexican War

1. Contract made and entered June 2, 1845, Louisville State of Kentucky, St. George Croghan and William Davis.

2. John Croghan to [Mrs.] Ann [Jesup, Washington City], December 17, 1845, LC.

3. John Croghan to Major General Jesup, April 28, 1846. LC.

4. Ellen Pearce Bodley [Vicksburg] to Judge William L. Bodley, May 9, 1946, TFHS.

5. Judge William L. Bodley to Ellen Bodley, May 15, 1846, TFHS.

6. James D. Livingston and Arthur Kelly, *A Livingston Genealogy*, The Friends Of Clermont, Sojourner Truth Library, NY, 1982.

7. John Croghan to Ann [Jesup], December 17, 1845, LC.

8. Geo. Croghan [Monterey] to John Croghan, October 5, 1847, LC.

9. George Croghan [Locust Grove] to Genl. Jesup, December 15, 1847, LC.

10. George Croghan [Baton Rouge] to Major General Jesup, November 4, 1848, LC.

11. Charles Richard Williams, *Address Delivered Before the George Croghan DAR Chapter,* August 1, 1903, OAHS.

12. Ibid.

13. George Croghan [Pasagonia, Miss.] to General Jesup, September 1, 1848, LC.

14. Francis Paul Prucha, *Army Life on the Western Frontier,* Introduction xxi, Norman: Univ. of Oklahoma.

15. Robert Selph Henry, *The Story Of The Mexican War,* 1950, Bobbs-Merrill Co. Inc., NY, p. 75.

16. George Croghan (Baton Rouge) to Genl. Jesup, November 4, 1848, LC.

17. *The Examiner,* Vol. 2, # 85, Jan. 27, 1849, p. [3], col. [2]. Copied from *New Orleans Picayune,* Jan. 9, 1849, TFHS.

18. George C. Gwathmey (Louisville) to General Jesup, January 18, 1849.

19. Silas Bent McKinley and Silas Bent, *Old Rough and Ready,*1946.

20. J. T. Taylor to John O'Fallon, March 20, 1849, MHS.

21. Croghan Papers, MSS and Archives Dept., NY Public Library.

Chapter Eight
Serena and Tinie on Their Own

1. Thomas Jesup [Washington City] to Mary Jesup Blair [California]. 1851, HL.

2. John R. Livingston, 26 September 1851, *New York Herald.*

3. George Croghan III, Sacramento Co. CA. to Albert C. Janin, Esq. Wash. D.C., May 1902, HL.

4. Eliza (McEvers) Livingston. February 17, 1848, *New York Herald.*

5. David Sever Lavender, *California,* 1976, Norton & Co., NY, p. 59.

6. James Blair to Mary J. Blair, September 30, 1853, HL.

7. HL, James Blair [Sacramento, CA] to Mary S. Blair, October 16, 1853.

8. Mrs. George Wheaton [Pacadena. CA] to Melzie Wilson, March 2, 1988.

9. Virginia Wheaton [Santa Barbara, CA] to Melzie Wilson [Lou. KY] May 2, 1990.

10. Ibid., September 14, 1990. (Received 18 page letter).

11. Ibid.

12. Ibid.

13. Ibid.

Chapter Nine
Civil War—Orphans—California

1. Alfred H. Guernsey and Henry M. Alden, *Harpers Pictorial History Of The Civil War,* 1866, Fairfax Press, p. 144. [Noted; November 1861].

2. Charles R. Williams, PH.D., *Address Delivered To George Croghan DAR Chapter,* Ausust 1, 1903, OAHS.

3. Jefferson County Deed Book, 221, p. 549.

PART III

Mary Croghan and the Englishman

Chapter One
O'Haras of Pittsburgh

1. Morrison Foster, *The Ancestry of Mrs. Captain Edward W.H. Schenley*, [Pittsburgh] The Historical. Soc. of Western PA, p. 17, University of Pittsburgh.

2. Ibid.

3. Ibid.

4. Ibid.

5. Ibid., p. 20.

6. The Historical Society of Western Pennsylvania Archives, Denny-O'Hara, MSS#51, Box 12, Folder 3. Pittsburgh, March 16, N.D.

7. Morrison Foster, *The Ancestry of Mrs. Captain Edward W.H.Schenley*, N.D. University of Pittsburgh, p. 20.

8. Margaret Pearson Bothwell, *Historical Society Notes,* Western Pennsylvania Historical Magazine, Vol. 47, No. 4 (1964), 366-367.

9. George Hancock to William Croghan, Jr., March 25, 1824, Darlington Memorial Library, University of PA.

10. William Croghan to Ann Jesup, August 9, 1835, LC.

11.Wm. Croghan, Jr. to Jesups, June 18, 1826, LC.

12. Ibid., August 1,1826.

13. Mary Schenley to cousin Libby [Pittsburgh], April 10, 1844, Margaret Scully Collection, Historical Society of Western, PA.

14. William Croghan, Jr. to Harmon [Brother-in-law], October 25, 1827, THS of WP.

15. William Croghan, Jr. to Jesups, November 22, 1827, LC.

16. William Croghan, Jr. to Wm. Croghan III, Fall 1827, THS of WP.

17. William Croghan, Jr. to Jesups, January 28, 1828, LC.

18. *The Western Pennsylvania Historical Magazine,* Vol.47, No. 3 (1964), 263–264.

19. James Craik, *Historical Sketches of Christ Church, Louisville,* Morton & Co. 1862, p. 31.

20. William Croghan, Jr. to Jesups, January 30, 1828, LC.

21. William Croghan, Jr. to Frederick Rapp, Jr. [Economy, PA], January 8, 1831. Darlington Memorial Library, Pittsburgh.

22. George Swetnam, *The Pittsburgh Press,* Sunday 25, 1954, p. 4.

23. Denny-O'Hara Family Papers, 1769–1949, The Historical Society of Western PA, Archives MSS# 51.

Chapter Two
Mary Croghan Had Two Families

1. Denny-O'Hara Family Papers, 1769–1949, The Historical Society of Western PA, Archives MSS# 51.

2. Mary E. Croghan to Lucy Ann Jesup, August 30, 1833, LC.

3. William Croghan to Charles W. Thruston, May 21,1832, TFHC.

4. William Croghan to Major Gen. Jesup, October 4, 1831, LC.

5. William Croghan to Ann Jesup, September 16, 1833, LC.

6. John Tucker Howard, *Stephen Foster, America's Troubadour,* NY, p. 21.

7. Sarah B. Smith, *A Tribute To Stephen Collins Foster,* 1990 GBA Printing, Bardstown, Ky. p. 8.

8. Ibid., p. 6.

9. *Pittsburgh Sun-Telegraph Pictorial Living,* February 15, 1959.

10. William Croghan to Charles W. Thruston, October 24, 1834, TFHC.

11. Mary Croghan to William Croghan, December 11, 1840, The Historical Society of Western PA.

12. *The Western PA Historical Magazine,* Vol. 9, 1926, p. 214.

13. Alberta McLean, J.P., New Zealand, May 29, 1946, Schenley Family, (AA Denny-O'Hara), University Of Pittsburgh.

Chapter Three
Romance That Rocked Two Continents

1. Mary Schenley [Paramaribo] to William Croghan, Pittsburgh, Septermber 3, 1842, Darlington Memorial Library, PA.

2. Henry Delafield to Wm. Croghan, February 8, 1842, Darlington Memorial Library, PA.

3. *Pittsburgh Sun-Telegraph,* September 1941.

Recovering from his first shock, Mr. Croghan appealed to the government in Washington to send out boats to intercept the vessel on which his sheltered young daughter had sailed with her bridegroom, old enough to be her Father. The government failed to find the honeymooners because wily Capt. Schenley had stopped enroute to England on an island, perhaps Bermuda, it is suggested in a yellowed clipping owned by the Carnegie Library Pennsylvania Room.

4. George Croghan to John O'Fallon, February 16, 1842, LC.

5. John Croghan to General. Jesup, May 9, 1842, LC.

6. *Pittsburgh Press,* May 29, 1951, A member of the Pennsylvania Legislature.

Chapter Four
Surinam—Dutch Guinea

1. Mary Schenley to William Croghan, September 1842, Darlington Memorial Library, Pittsburgh.

2. Edward Schenley to Elizabeth Denny, July 9, 1843, The Historical Society of Western PA, Archives MSS#51.

3. Ibid., July 9, 1843.

4. Mary Schenley to Elizabeth "Mother" Denny, August 7, 1843, Darlington Memorial Library, Pittsburgh.

5. Ibid.

6. Scully Collection, Pittsburgh, THS of WP, April 13, 1844, MSS#34.

7. Mary Schenley to Sister, Mar. 2, 1846, Pittsburgh, Denny-O'Hara MSS#51, THS of WP.

8. Mary Schenley to Libby Denny, Jan. 28, 1846. Pittsburgh, THS of WP.

9. William Croghan to Elizabeth Denny, Oct. 1, 1845, Pittsburgh, Denny Family, MSS#51, THS of WP.

10. Ibid.

11. William Croghan, (Southampton) to Elizabeth Denny, March. 2, 1846, Denny Family, MSS#51, THS of WP.

12. Ibid.

13. Ibid.

14. Ibid.

15. William Croghan to Harmar Denny, May 31, 1847, London, THS of WP.

16. Ibid.

17. William Croghan to Elizabeth Denny, Oct. 11, 1848, Pittsburgh, THS of WP.

Chapter Five
Schenley Children

1. Marie M. Swigan, *Stanton Heights Golf Link Is Shrine Of City's History*, c1930, *The Pittsburgh Press*, THS of WP.

2. Gilbert Love, *Schenley-A Mighty Name*, February 23, 1934, *The Pittsburgh Press*, THS of WP.

3. Murray Bayler, *Picnic House Recalls Mary Schenley's Elopement*, Nov. 1, 1945, *Pittsburgh Post-Gazette*.

4. George Swetnam, *A New Chapter of An Old Romance*, Apr. 25, 1954, *The Pittsburgh Press*, p. 6. THS of WP.

5. Edward W. H. Schenley to Mrs. Denny (Elizabeth) Jan. 4, 1862, [Letter has a black border], Denny-O'Hara MSS, THS of WP.

6. Mary Schenley to Elizabeth Denny, Oct. 3, [1862], Denny-O'Hara MSS, THS of WP.

7. Ibid., Dec. 6th, 1862.

Chapter Six
Mary Wins Over U.S. Courts

1. Morrison Foster, *The Ancestry of Mrs. Captain Edward W.H. Schenley*, (N.D.), pp. 21–23, Foster Hall Collection, Stephen Foster Memorial University of Pittsburgh.

2. Gilbert Love, *Schenley—A Mighty Name*, Feb. 23, 1934, *Pittsburgh Press*, p. 25.

3. Mrs. Alberta McLean, J.P. May 29, 1949, New Zealand, Denny-O'Hara MSS, THS of WP.

4. Ibid.

5. Western Pennsylvania Historical Magazine, *One Hundredth Anniversary of the Birth of Mrs. Schenley,* Vol. 9, 1976, p. 216.

6. Ibid.

7. Will Book No. 77, p. 359, records of Allegheny County, Pa.

8. Western Pennsylvania Historical Magazine, *One Hundredth Anniversary of the Birth of Mrs. Schenley,* Vol. 9, 1976, p. 218.

9. Hermione Ellenborough to Julia Jesup, November 1905, Janin Family Collection, HL.

10. Ibid., 1910 Wimblesham Court, Surrey England.

11. Hermoine Ellenbourgh to Violet Janin, May 15, 1915, HL.

12. *Exhibition and Sale, The Mary Schenley Collection Victorian Furniture,* at Kaufmann's, [Eleventh Floor], Pittsburgh. 1931, THS of WP.

13. Bernice Shine, *Schenley Park Donated by Girl Whose Romance Shocked a Queen,* Sept. 15, 1941, *Pittsburgh Sun-Telegraph.*

14. Mrs. Alberta McLean, J.P. to Charles W. Shetler, Lecturer of History, University of Pittsburgh, May 29, 1949, Denny-O'Hara MSS, THS of WP.

15. *Pittsburgh Sun-Telegraph Pictorial Living,* "Beauty out of the past," Feb. 15, 1959, THS of WP.

PART IV
Widow of Lafayette Square

Chapter One
At Home in Kentucky

1. Mrs M.L. Clark, May 30, 1901. Dye Collection.

2. Ibid.

3. George Hancock (Louisville) to Mary Jesup Blair, May 30, 1865, HL.

4. Frances Preston Blair to "Dear Madame" [Ann Croghan Jesup], Fall – 1845, HL.

5. Blair and Lee Family Papers, (1818–1906), Princeton University, NJ, Box 249 (Folder 3).

Elizabeth Blair was a great favorite of Gen. Jackson who insisted she live at the White House one winter until the cellar of the newly acquired Blair house could be drained because of poor health.

6. Thomas Jesup assumed command May 20, 1836, in December, he took command of the Seminole War in Florida. Jesup had been part of Harrison's "scandal machine"and never let up. The following is found in John O'Fallon Papers, (MHS); Jesup used Honorable Joseph M. White to have Scott removed so that Jesup would be assigned to the command. In Jesup's mind, he, himself was the true leader. General Jackson directed that an order be issued commanding General Scott, to withdraw from the command in Florida. Order was issued, bearing date June 1st 1836. See page 476 – Vol. 3rd Executive Documents of the 25th Congress 2nd Session. This order was issued in consequence of a letter recd by the President, from Hon.

Joseph M. White. (Delegate from the Territory of Florida) bearing date the 28th May 1836 & is endorsed by the President as follows: "A copy of this letter to be sent to General Scott, with an order to withdraw from the command in Florida." General Jesup's letter (private letter) to his friend F.P. Blair was written June 20th 1836. When Mr. White's letter was written, Genl. Jesup was on the road to assume commandin the Creek country.

7. LC, William Croghan (Pitts.) to Mrs. Ann H. Jesup (Washington City D.C.), October 25, 1835.

8. LC, John Croghan to Maj. Genl. T.S. Jesup (Tampa Bay, Florida) January 22, 1837.

9. Ibid.

10. Ibid.

11. Uncle John Croghan to William Jesup, January 16, 1846, LC.

12. Ibid.

13. Lucy Jesup Sitgreaves (Chestnut Hill, Boston) to Mary Jesup Blair, December 9, 1905, HL.

14. John Croghan to Major General. T. S. Jesup, [Tampa Bay, Florida], May 20, 1837, HL.

15. Ibid.

16. George Croghan [Washington] to John and Ann at Locust Grove, April 17, 1838, HL.

17. Floyd B. Largent, Jr., "The Florida Quagmire," *American History*, October 1999, p. 43.

18. Alfred Jackson Hanna, *Lake Okeechobee*, The Bobbs-Merrill Co. NY. p. 44.

19. Floyd B. Largent, Jr., "The Florida Quagmire," *American History*, October 1999, p. 45.

20. Ibid.

21. Alvin M. Josephy, Jr., *The Patriot Chiefs*, 1961, Viking Press, NY, p. 205. Alfred Jackson Hanna, *Lake Okeechobee*, The Bobbs-Merrill Co. NY, p. 45.

22. Floyd B. Largent, Jr., "The Florida Quagmire," *American History*, October 1999, p. 45.

23. Ibid.

24. Jesup's conduct provoked an inquirey—Thomas H. Benton of Mo. intervened and stopped it.

Jesup had become a favorite of President Monroe during the War of 1812. The New England States bitterly opposed the War of 1812 and in 1814, leaders in Rhode Island, Connecticut, Massachusetts and parts of New Hampshire proposed a convention and had secret sessions for three weeks. They proposed greater independence to the individual states. Opponents charged that the Hartford Convention had plotted secession from the Union, [*The World Book Encyclopedia*, Vol. 8, Field Enterprises, Inc. Chicago, p. 3296].

Monroe, the Secretary of War was concerned and needed a spy to be sent to Hartford, but it should be someone from New Jersey pretending to recruit for a regiment, but to watch and report. Discovering a Jesup from NJ had fought in the Revolutionary War , and realizing they had a Jesup enlisted in the army, contacted Thomas Jesup. He was sent to Harford, but news of the war ending, nothing happened. However, this one act opened doors for Jesup.

Chapter Two
Beaus and Belles of Washington

1. William Croghan [Pic Nic] to Ann H. Jesup, Washington City, July 5, 1835, LC.

2. Amy La Follette Jenson, *The White House And Thirty Four Families* (New York: McGraw-Hill, 1965), p. 24.

3. Major Gist Blair, *Lafayette Square, Reminiscences Of The District Of Columbia of Washington, 79 years ago*, edited by Sarah E. Vedder, p. 17. Blair-Lee Family Papers, Princeton University, NJ.

4. William Seale, *The Presidents House*, Vol. I (Washington, D.C.: White House Historical Association, 1986), p. 229.

5. Ibid.

6. John Frederick Dorman, *The Prestons Of Smithfield And Greenfield In Virginia*, The Filson Club Publications, p. 341.

7. 29. Boynton Merrill, Jr., *Jefferson's Nephews* (Princeton: Princeton University Press, 1976), p. 184.

8. Ibid.

9. Blair-Lee Family Papers, Elizabeth Blair Lee Retold by Blair Lee, (1857–1944), Box (249), Folder (3). Princeton University, NJ.

10. George Hancock (Louisville) to John Croghan, Saturday June (?) 1845, LC.

11. Francis Blair to Ann Croghan Jesup, Fall 1845, HL.

12. 29. Boynton Merrill, Jr., *Jefferson's Nephews* (Princeton: Princeton University Press, 1976), p. 185.

13. Thomas and Ann Croghan Jesup Family Chart, (Deaths), April 24, 1846, HL.

14. James Blair (Pic Nic, Pittsburgh) to Mary J. Blair, September 18, 1846, HL.

15. Ibid.

16. Ibid.

17. Ibid., February 25, 1847.

18. Ibid., February 27, 1847.

19. Ibid.

20. Ibid., March 23, 1847.

21. Ibid., [US Steamer Polk] March 24, 1847.

22. Ibid., March 31, 1847.

23. Ibid., [South Hampton, England] June 15, 1847.

24. Ibid., [London] July 1, 1847.

25. Ibid., [American Hotel, New York] September, 1847.

26. Ibid., [American Hotel, New York] March, 1848.

27. Ibid., November 19, 1848.

Chapter Three
California

1. James Blair (Steamer Falcon) to Mary Jesup Blair, February 2, 1849, HL.

2. Ibid., February 2, 1849.

3. Ibid., March 24, 1849.

4. James Blair (San Francisco) to Mary Jesup Blair, April 6, 1849, HL.

5. Ibid.

6. Ibid., March 24, 1849.

7. Ibid., April 30, 1849.

8. Ibid.

9. Ibid., June 27, 1849.

10. Ibid.

11. Ibid., June 29, 1849.

12. Ibid., July 23, 1849.

13. Ibid., September 26, 1849.

14. Ibid.

15. Ibid.

16. Ibid.

17. Ibid., February 29, 1850.

18. Ibid., June 30, 1850.

19. Ibid., October 15, 1850.

20. James Blair to Mary J. Blair, November 15, 1850. HL.

21. Major Jesup (Washington City) to Mary J. Blair, February 8, 1851, HL.

22. Mary J. Blair (San Francisco) to Eliza Blair Lee, December 15, 1851, HL.

23. Ibid., February 15, 1852.

24. Ibid., March 1, 1852.

25. James Blair to Mary J. Blair, December 1, 1852, HL.

26. Ibid., December 20. 1852.

27. Ibid., July 31, 1853.

28. Christopher Wyatt (San Francisco) to General Jesup (Washington), December 16, 1853, HL.

29. F.P. Blair to Mary Blair, January 1854, HL.

30. Ibid.

31. Ibid.

32. James Blair (California) to Mary Jesup Blair (Washington D.C.) October, 1949, HL.

Chapter Four
Widow of Lafayette Square

1. Mary Blair (Moorings) to General Jesup, (No date) HL.

2. William Jesup (Spring Grove) to Janie Jesup Nicholson, April 25, 1857. HL.William was b. 27 June, 1833; admitted to West Point Military Academy, cadet at large, 1850; d. 17 November, 1860.

3. Charles Jesup (Spring Grove) to "Dear Mamie" (Mary J. Blair), March 6, 1861, HL. Charles

Edward, B. 14 March, 1835; admitted to West Point, cadet at large, 1850; Brevet 2nd Lieut. 10th Infantry, 1 July, 1858; 2d Lt. 6th Infantry, 31 May 1859; resigned 20th Aug., 1860; D. 22 April, 1861. Jesup Genealogy, Ingersoll's, *History of the War Department.*

4. Jesup Family Records, HL.

5. C. E. Jesup to "My dearest Mary," April 22, 1861, HL.

6. Rabbi David Philipson, *Letters of Rebecca Gratz* (Philadelphia: The Jewish Publication Society of America, 1929). p. 422.

7. Roger Brooke Farquhar, *Historic Montgomery County, Maryland, Old Homes and History,* (Louisville, National SAR Library), p. 31.

8. Elizabeth Blair Lee (1818–1906) as told to son, Blair Lee, Blair and Lee Family Papers, Princeton Univ. Lib.

9. HL, Violet Blair Diary, May 25, 1868.

10. Janin Family Collection, Huntington Lib. San Marian, California, Papers (7), Louis Janin [1803–1874] was born and educated in Vienna, emigrated to the U.S. in 1828, established residence in New Orleans, was naturalized, admitted to the bar, opened law offices, invested in sugar plantations. He married Juliet Covington. Six sons reached maturity and were educated mainly in Europe. Three sons engaged in mining in California, Nevada, Japan, and elsewhere. One son killed in Civil War, Manassas in 1862, another engaged in business in Minnesota and California. Albert practiced law and entered into politics in New Orleans. He managed Mammoth Cave and was the first to make it a successful enterprise.

11. Violet Blair Janin (Bowling Green) to Mary J. Blair, November 8, [1874], HL.

12. Blair Lee, Attorney At Law, Washington, D.C., to Mrs. Mary J. Blair, August 18, 1900, HL.

13. Albert Janin (Lafayette Sq.) to (Julia Jesup) in Europe. April 2, 1902, HL.

> My Dear Aunt Julia,
>
> How I wish you were here. An awful bereavement has come upon us. Jesup, the life and soul of the household; the sweetest tempered, Kindest-heared, most courteous gentleman I ever knew, a son and brother who made earthly idols of his mother and sister, and who was to me the devoted friend I ever had, is dead. He had just come in from a walk and sat down to the lunch table, when, after a few jovular remarks, a sudden pallor came over his face, he reeled slightly, his mother rushed to his side, his head reclined for a moment on her breast and he fell to the floor dead of cerebrel hemorrhage.

14. Lucy Jesup Sitgreaves (Woodstock, VT) to Mary J. Blair, August 13th, 1903, HL.

15. William Hancock Clark to Mary Jesup, 1903, Dye Collection.

PART V

The Croghan Heirs of Mammoth Cave

1. Cecil E. Goode, *World Wonder Saved*, The Mammoth Cave National Park Association, Ky. 1925, p. 49.

2. Ibid., p. 8.

3. H. C. Hovey, *Gleason's Pictorial Drawing-Room Companion*, June 5, 1852 pp. 356–357, Ekstrom-Belknap Campus, Louisville. KY.

4. Roger Brucker & Richard A. Watson, *The Longest Cave* (New York: Alfred A. Knopf, Inc.,

1976, p. 269.

5. Cecil E. Goode, *World Wonder Saved*, The Mammoth Cave National Park Association, 1925, Ky. p. 6.

6. Ibid., p. 10.

7. Roger Brucker & Richard A. Watson, *The Longest Cave* (New York: Alfred A. Knopf, Inc., 1976, p. 269.

8. Ibid., p. 271.

9. The Filson Club History Quarterly, *Groping For Health In Mammoth Cave* Vol. 20, 1946, p. 302.

10. John Croghan's Will; Jefferson County Probate: February 5, 1849. The three trustees were George Gwathmey (died very soon), William F. Bullock didn't accept, Judge Underwood resigned in 1870, William R. Thompson succeeded Underwood, but died very soon. Another provision of the will gave Locust Grove slaves (except Isaac) to the same three trustees, allowing George Croghan (brother) to occupy and use Locust Grove during his lifetime, then to Nephew St. George (usually called George). The slaves were to be hired out for four (4) years and after that time, they were to be hired out for three (3) years "to prepare for freedom". After three (3) years, the slaves were to be emancipated with all their increase. Isaac was emancipated immediately.

11. Judge J. R. Underwood Mammoth Cave Report, February 13, 1868, HL.

12. Christopher B. Wyatt [Westchester, NY] to Jesup Blair [Washington D.C.], November 23, 1876, HL.

13. Christopher B. Wyatt (New York) to Charles Thruston, Esqr., [Louisville, KY], May 22, 1852, HL.

14. Christopher B. Wyatt [NY] to Major A.S. Nicholson [Washington D.C.], February 3, 1876, HL.

15. Byron Renfro, Attorney At Law, Edmondson Circuit Court, KY, 1870.

16. Christopher. B. Wyatt [Westchester, NY] to Jesup Blair [Washington D.C.], October 2, 1876, HL.

17. Ibid., October 2, 1876.

18. Ibid., November 23, 1876.

19. Ibid., October 16, 1876.

20. George Croghan [Mammoth Cave Hotel] to Augustus Nicholson, June 22, 1880, HL.

21. Ibid., September 6, 1880. George Croghan learned that Nicholson had written Miller, requesting his report of the situation at Mammoth Cave. Nicholson was deceiving George by covering up his own reports sent to Miller and Wyatt. George received notice that he was to leave Mammoth Cave premises at once. Augustus Rodgers (California) also, found Nicholsons's letters convincing.

22. HL, Augustus F. Rodgers to Augustus Nicholson, December 24, 1880, HL.
Immediately after George Croghan was forced to leave the premises, the Jesup women invested a great amount of money into the Cave. Rodgers referred to the Cave as Jesup property, but reiterated the fact: "I suppose we are in duty bound to make the change you suggest, but it will make no change in the income devisable from the property. No change to the heirs, unless the lease." There is no evidence found by this writer that George ever revealed his findings in the 'books' and then was threatened at gun point that he was never to reveal to anyone about his discoveries. Years later when summoned to Kentucky as a witness in Edmonston Court to testify as to what he had discovered while serving as Clerk at the Cave. The local Kentucky court dismissed the charges

23. Augustus Rodgers to Augustus Nicholson, December 24, 1880, HL.

24. Rodgers to Julia Jesup, July 12, 1884, HL.

25. Ibid., September 21, 1884.

26. Alvin F. Harlow, *Weep No More My Lady* (New York: McGraw-Hill Book Co., Inc., pp. 268–269.

27. George Croghan III, [Sacramento Co. California] to Honorable John M. Wilkins, Bowling Green, KY, April 24, 190l, HL.

28. Albert Janin [Mammoth Cave, KY] to Aug. F. Rodgers, April 26, 1901, HL.

29. Ibid., [No Date]

30. Augustus Rodgers to Mrs. James Blair, My 16, 1902, HL.

31. Ibid.

32. Lucy Croghan Browne to Violet Blair Janin, January 3, 1915, HL.

33. H. C. Hovey, *Gleason's Pictorial Drawing-Room Companion* (Louisville: Ekstrom-Belknap Campus, 1909)

34. Cecil E. Goode, *World Wonder Saved*, The Mammoth Cave National Park Association, KY, 1925, p. 21.

35. Cornelia Nokes to Albert Janin, November 24, 1921, HL.

36. Ibid., September 5, 1923.

37. Ibid., January 8, 1924.

38. Lucy Browne to Violet Janin, September 22, 1924, HL.

39. Cornelia Nokes to Violet Janin, April 24, 1824, HL.

40. Ibid.

41. Cecil E. Goode, *World Wonder Saved*, The Mammoth Cave National Park Association, KY, 1925, p. 22.

42. Lucy Browne to Violet Janin, 1911, HL.

43. Spencer Browne (New Mexico) to Miss Mary Sitgreaves (Chestnut Hill, MA) September 4, 1924, HL.

44. Ibid.

45. Ibid.

46. Spencer Browne to Violet Janin, April 29, 1925, HL.

47. Cecil E. Goode, *World Wonder Saved,*, Mammoth Cave National Park Association, p. 25.

48. Ibid., p. 33.

49. Roger W. Brucker & Richard A Watson, *The Longest Cave*, 1976, Alfred A. Knopf, NY, p. 285.

50. Cecil E. Goode, *World Wonder Saved*, Mammoth Cave National Park Association, p. 19.

BIBLIOGRAPHY

Abbott, Carl, *The Great Extravaganza, (1905) Portland And The Lewis And Clark Exposition.*
[Revised Edition], Oregon Historical Society.

Abernethy, Thomas Perkins, *The Burr Conspiracy.* New York: Oxford Press, 1954.

Ambler, Charles Henry, *A History of Transportation In The Ohio Valley.* Glendale, Calif.: A.H.
Clark Co., 1932.

Ambrose, Stephen E., *Undaunted Courage.* New York: Simon & Schuster, 1996.

Bacon, Edgar Mayhew, *Chronicles of Tarrytown and Sleepy Hollow.* New York: G. P. Putnam's
Sons, 1897.

Bakeless, John, *Back Ground To Glory.* New York: Lippincott, 1957.

Bakeless, John, *Lewis and Clark, Partners in Discovery.* New York: Morrow and Co., 1947.

Baldwin, Leland D., *The Keelboat Age On Western Waters.* Pittsburgh: University of Pittsburgh
Press, 1941.

Battle-Perrin-Kniffin, *Kentucky, A History of the State.* Louisville: Battey Pub. Co., 1885.

Bent, Silas and Silas Bent McKinley, *Old Rough and Ready.* New York: Vanguard Press.

Bertram Wyatt-Brown, *Southern Honor.* New York: Oxford Press, 1982.

Betts, Robert B., *In Search of York.* Boulder: Colorado Associated University Press, 1985.

Biddle, Nicholas, Edited: *The History of the Expedition Under the Commands of
Captains Lewis and Clark.* 1814.

Billington, Ray Allen, *Western Expansion, A History Of The American Frontier.* New York:
Macmillan Co., 1959.

Boardman, Fon W. Jr., *America and the Virginia Dynasty 1800–1825.* Printed in the U.S., 1974.

Bodley, Temple, *George Rogers Clark—His Life And Public Services.* Cambridge: Riverside
Press, 1926.

Boorstin, Daniel J., *The Americans.* New York: Random House.

Briney, Melville O., "Fond Recollection, Sketches Of Old Louisville," *Louisville Times,* KY.

Cassaday, Ben, *The History Of Louisville.* Louisville: Hull and Brother, 1852.

Christian, Archer and Susanne Massie, Editors, *Homes and Gardens in Old Virginia.* Garden
Club of VA.

Clark, Daniel, *Gen. James Wilkinson.* Philadelphia: Hall & Pierie, 1809.

Clark and Edmonds, *Sacagawea of the Lewis and Clark Expedition.* Los Angeles, 1979.

Cleaves, Freeman, *Old Tippecanoe.* Connecticut: American Political Press, 1996.

Coit, Margaret L., *The Growing Years,* New York: Time-Life Books.

Craik, James, *Historical Sketches of Christ Church.* Louisville: John P. Morton, 1862.

Crowder, Lola Frazer, *Early Louisville, Kentucky Newspaper Abstracts, 1806–1828.*
Galveston: Frontier Press, 1995.

Dangerfield, George, *Chancellor Robert R. Livingston of New York.* New York: Harcourt, Brace
and Co., 1960.

Davis, Burke, *A Williamsburg Galaxy.* New York: Holt, Rinehart & Winston, 1968.

Dillon, Richard, *Meriwether Lewis, A Biography.* New York: Coward-McCann, Inc., 1965.

Dollarhide, William, *American Migration Routes 1735–1815.*

Donovan, Frank, *The Women In Their Lives, The Distaff Side Of The Founding Fathers.* New
York: Dodd, Mead, 1966.

Dupuy, Ernest and Trevor, *An Outline History of the American Revolution.* New York: Harper
& Rowe.

Durrett, Reuben T., *The Centenary of Louisville.* Louisville.: J.P. Morton & Co., 1893.

English, William H., *Conquest of the Northwest.* Indianapolis: 1896.

Evans, Elizabeth, *Weathering The Storm, Women of the American Revolution.* New York:
Scribners.

Every, Dale Van, *Forth to the Wilderness, The First American Frontier.* New York: Wm. Morrow & Co., 1961.

Floyd, William Barrow, Jouett-Bush-Frazer, *Early Kentucky Artists.* Lexington, Ky.: Published by Author, 1968.

Ford, Alice, *Audubon, By Himself.* New York: The Natural History Press.

Graff, H. and John A. Krout, *The Adventure of the American People.* Skokie, Ill.: Rand McNally & Co., 1971.

Gwathmey, John H. *Twelve Virginia Counties.* Richmond, Va.: The Dietz Press, 1937.

Goebel, Dorothy Burne, PH.D., *William Henry Harrison.* Indiana Library and Historical Dept., 1926.

Hanna, Alfred Jackson and Kathryn Abbey Hanna, *Lake Okeechobee.* New York: Bobbs-Merrill Co., Inc.

Harris, Bill, *Home Of the Presidents.* New York: Crescent Books, 1987.

Hebard, Grace Raymond, *Sacajawea.* New York: Dover Publications Inc., 1979.

Hendrickson, Charles Cyril, *George Washington, A Biography in Social Dance.* Connecticut: The Hendrickson Group.

Henry, Robert Selph, *The Story of the Mexican War.* New York: Bobbs-Merrill Co., Inc.

Holmberg, James J., *Dear Brothe.,* Published in Association with TFHS, Princeton: Yale University Press, 2002.

Holmes, Oliver W. and Peter T. Rohrback, *Stagecoach East.* Washington, D.C.: Smithsonian Institution Press, 1983.

Irvin, Helen Deiss, *Women in Kentucky.* Lexington, Ky.: University of Kentucky Press, 1979.

Jackson, Donald, Edited the letters of *Lewis and Clark Expedition.* Urbana: University of Illinois Press, 1962.

Jackson, Donald, *Thomas Jefferson & The Stony Mountains.* Norman: University of Oklahoma Press.

James, James Alton, *George Rogers Clark.* Chicago, 1928.

Josephy, Alvin M. Jr., *The Patriot Chiefs.* New York: The Viking Press, 1961.

Kukla, Jon, *A Wilderness So Immense.* New York: Random House, Inc., 2003.

Laas, Virginia Jeans, *Love and Power in the Nineteenth Century.* City: University of Arkansas Press, 1998.

Largent, Floyd B. Jr., "The Florida Quagmire," *American History,* October 1999.

Lavender, David, *The Way To the Western Sea.* New York: Harper & Rowe, 1817.

Lewis, Mary Newton, "A Postscript to In Search of York, *We Proceeded On,* 1990.

Livingston, Kelly, Brandt, *A Livingston Genealogy,* 1986, Co-sponsored by The Order of Colonial Lords of Manors in America, for The Friends of Clermont, NY.

Lavender, David, *California.* New York: W.W. Norton & Co., 1976.

Lorant, Stefan, *Pittsburgh.* New York: Doubleday & Co., Inc., 1964.

Malone, Dumas, *Jefferson the President.* New York: Little, Brown and Co., 1974.

Malone, Dumas, *Thomas Jefferson's Monticello.* Charlottsville, N.C.: Thomasson-Grant, 1983.

McAfee, Robert B., *History of the Late War.* 1816.

McClanahan, William A., *Magruder's American Government.* Boston: Allyn and Bacon, Inc. 1966.

McCullough, David, *John Adams.* New York: Simon & Schuster, 2001.

McMeeken, Isabel, *Louisville, The Gateway City.* New York: Julian Messner, Inc., 1946.

McMurtrie, Henry, *Sketches Of Louisville.* Louisville: 1819.

Miller, John C., *The Colonial Image.* New York: George Braziller, 1962.

Morgan, Edmund S., *Virginians at Home, Family Life In The Eighteenth Century,* Williamsburg, Vir.: 1977.

Pirtle, Alfred, *Mulberry Hill,* The Register of The Ky. St. His. Society, Frankfort, KY. Vol.15, No. 43.

Prucha, Paul Francis, *Army Life On The Western Frontier,* [Official Reports Made Between 1826–1845 by Colonel George Croghan].

Quisenberry, Anderson Chenault, *Kentucky In The War Of 1812.* Clearfield Co. Inc. By Genealogical Publishing Co., Reprinted in 1989.

Rothery, Agnes, *Houses Virginians Have Loved.* New York: Bananza Books.

Rothery, Agnes, *Virginia, The New Dominion.* New York: Appleton-Century, 1940.

Rouse, Parke, Jr., *Planters and Pioneers.* New York: Hastings House.

Rouse, Parke, Jr., *The Great Wagon Road.* Printed in the U.S., 1973.

Saunders, Col. James Edmonds, *Early Settlers Of Alabama.* New Orleans: Genealogical Pub. Co. Inc., 1982.

Schmit, Patricia Brady, *Nelly Custis Lewis's Housekeeping Book.* Printed in U.S., 1810.

Seale, William, *The President's House, Vol. I—Vol. II,* 1986, White House Historical Association.

Skeen, C. Edward, *John Armstrong, Jr. (1758–1843).* New York: Syracuse University Press, 1981.

Steffen, Jerome O., *William Clark: Jeffersonian Man on the Frontier.* Norman: University of Oklahoma Press, 1977.

Taylor, Dale, *Everyday Life in Colonial America.* Cincinnati: F & F Publications, 1997.

Thwaites, Reuben Gold, *Journals of the Lewis and Clark Expedition, 1804–1806.* New York.

Tischendorf, Alfred and E. Taylor Parks, Editors, *Diary and Journal of Richard Clough Anderson,* Jr. Durham: Duke University Press, 1964.

Todd, Lewis Paul and Merle Curti, *Rise Of The American Nation.* New York: Harcourt, Brace & World, 1966.

Volwiler, Albert T., *George Croghan and the Westward Movement (1741–1782).* Cleveland: Arthur H. Clark Co., 1926.

Wade, Richard, *The Urban Frontier.* Urbana, Ill.: University of Chicago Press.

Waldrup, Carole Chandler, *President Wives.* Jefferson, N.C.: McFarland & Co. 1925.

Walker, William, *Southern Harmony and Musical Companion.* Lexington, Ky.: University of Kentucky Press, 1987.

Waterman, Thomas T., *Mansions of Virginia.* Chapel Hill: University of North Carolina Press, 1946.

Wheeler, Olin, *The Trail of Lewis and Clark (1804–1904),* Vol. I. New York: Putnam, 1904.

Whitton, Mary Ormsbee, *First First Ladies 1789–1865.* New York: Hastings House.

Williams, Charles Richard, *Major George Croghan,* Columbus, Ohio: Ohio State Historical Society, 1903.

Wilson, Charles Reagan, & William Ferris, *Encyclopedia of Southern Culture.* Chapel Hill: Univsity of North Carolina Press,

Wilstack, Paul, *Tidewater Virginia.* Indianapolis: Bobbs Merrill Co., 1929.

Wyatt-Brown, *Southern Honor.* New York: Oxford University Press, 1982.

Zimmerman, Catharine S., *The Bride's Book.*